W0037041

Breach at Vosmeer, one of the many breaches caused by the flood of February 1st, 1953
Vosmeer is the place of origin of the Roosevelt family

Recently published: the fifth enlarged edition of

Dredge
Drain
Reclaim

THE ART OF A NATION

BY

Dr. JOHAN VAN VEEN

Late Chief Engineer Rijkswaterstaat

1962. 200 pages. With 112 figures, photographs and maps, 27 of which coloured. 4to.
Cloth Guilders 22.50

One guilder = ab. $ 0.278 = ab. 2 sh = env. N.F. 1,36 = ca. DMW. 1.10

MARTINUS NIJHOFF — PUBLISHER — THE HAGUE

This book contains a wealth of information historical as well as technological; it gives a clear picture of the Dutch people's continual fight against the sea. Dr. van Veen describes how The Netherlands have been wrested from the water by hard toil and at the cost of many human lives. There is no other country to be found in the world that has been moulded to such an extent by the hands of man: the making of hillocks in the early days of history; the construction of sea-walls or dikes and the digging of ditches and canals in modern times.

A large section is devoted to the Zuyderzee land reclamation project creating an additional 550.000 acres of fertile land and providing a fresh-water reservoir in the heart of the country. The final chapter deals with the flood disaster of February 1953 and the new gigantic Delta plan entailing the closing of the sea-arms in the south-western part of the country; this project involves the construction of four main barrier dams to a total length of 25 miles to be completed in 25 years. Thereby the Netherlands coastline will be shortened by no less than 435 miles.

CONTENTS: **Introduction.** Preface by Dr. J. A. Ringers – Foreword by Dr. Joh. van Veen – In honour of a master of the floods, by H. J. Stuvel. Chapter I: **Spade Work.** Beginnings – The 'Golden Hoop' – Perseverance – Success – Windmills – Ancient Dutch Abroad – Balance of Losses and Gains – Three Williams the First. Chapter II: **Dredges' Work.** 'Waterstaat' and 'Waterschappen' – A Kingdom for a Dredge! – The Spade once more – The Birth of the Dredge – The Rotterdam Waterway – The Amsterdam Waterway – Improvement of the Rivers – Canals for Inland Shipping. Chapter III: **Masters of the Floods.** Reclamation of the Zuiderzee – Walcheren – Scientific Investigation – Modern Dutch Abroad – The Tools – Riches. Chapter IV: **A new Storm, a new Start.** Dutch Floods Abnormal – The Vulnerable Country – Great Safety Project Ready – Engineers and Fleet on the Alert – Vierlingh's View – The Delta Plan – The Road Ahead – Luctor et Emergo – Our forgotten six Commandments.

Obtainable through any bookseller or direct from the publisher

DREDGE, DRAIN, RECLAIM

Dredge
Drain
Reclaim

THE ART OF A NATION

BY

Dr. JOHAN VAN VEEN

LATE CHIEF ENGINEER RIJKSWATERSTAAT

FIFTH EDITION

Springer-Science+Business Media, B.V.

1962

ISBN 978-94-015-1669-3   ISBN 978-94-015-2808-5 (eBook)
DOI 10.1007/978-94-015-2808-5

Copyright 1962 by Springer Science+Business Media Dordrecht
Originally published by Martinus Nijhoff, The Hauge, The Netherlands in 1962
Softcover reprint of the hardcover 5th edition 1962
All rights reserved, including the right to translate or to reproduce this book or parts thereof in any form.

CONTENTS

Chapter IV: A new Storm, a new Start

INTRODUCTION

During the occupation of the Netherlands the Germans made it impossible to carry out any maintenance work on our shores or any sounding, soil investigation or current-measurement work off the coast, in the estuary of the Scheldt or in the channels between the Frisian Islands. The work of Dr. Johan van Veen, then leader of this survey, therefore came to a standstill. He then came to me and asked me to give him some task, so that he, an indefatigable worker, could continue to have work, the best antidote against the German poison, which affected only permanently unemployed men.

I knew his love for the history of our traditional handling of the defence against the water. An all-round study had never been published, for in normal times a man with full knowledge of this type of work cannot find time for such a study, as water is our everlasting enemy, which must be kept under continual close observation.

From Dr. van Veen's book it will be clear that the Dutch manner of dredging, draining and reclaiming is a combination of traditions inherited from our ancestors and applied science to cope with modern demands.

This tradition is in our blood. A more intimate knowledge of it will, I hope, furnish a key to some of the salient points in our national character.

DR. J. A. RINGERS
Ex-Minister of Waterstaat

When *Dr. Ringers*, then Minister of Waterstaat, asked me to write this book I realized its necessity but felt uneasy about it. It would have to be an attempt to outline the age-long development of Dutch hydraulic engineering, and this would mean an historical rather than a technical study.

However, history was found to be interesting when one left out of account the battles and dates one learns at school and chose the incessant struggle of a nation for one's subject. My omission of detail has been necessary and intentional. The subject is worthy of more extensive treatment, but that would have been detrimental to the main object of interesting the men we wanted to reach.

In this retrospective mood I wish to pay special tribute to those, living or dead, who have shown any initiative in our country's endeavour to make something out of almost nothing. This historical study has made me appreciate that quality more than I did before. At the same time I want to thank those of my British friends who polished up my English – *Stephenson* of the Hughes Marine Instruments and *Forster* of the Royal Dutch-Shell Group have both shown much interest and sympathy. *Corlett* of the Ribble Estuary devised the title of the book, and some other English friends gave good advice. Likewise my thanks are due to 'Waterstaat' for providing various data. Last, but not least, I must express my appreciation of the initiative shown by *Mr. F. Kerdijk*, who – as printer and publisher – has launched this work. DREDGE, DRAIN, RECLAIM! May this book help to reveal the peculiar mission of my fellow-countrymen.

Some new items have been added to the later editions. Since the first edition in 1948 the Soil Charting Office made the discovery of the tremendous marine erosion of the sea marshes between 300 and 800 and allowed the chart page 23 to be printed. Further I found the genealogy of the baby in the cradle, who appears to have actually existed, and might even be the ancestress of most of the dredging, shipbuilding, and towing families of Holland. Two new chapters have been added for the fourth edition, one about the 'Three Williams I', the other about the 1953-flood. Being too busy myself I was much pleased to find that *Dr. Cassandra*, an expert in dike affairs, who had warned since 1937 that the southwestern dikes of Holland were too low – and who consequently had obtained this surname – was found willing to add the last of these two chapters to my book, and make the rest of it up to date.

'Cassandra' by the way, was the name of the princess who saw the insufficient strength of the defence of Troy, but could not prevent a calamity. The modest author would not have published his real name.

Dr JOH. VAN VEEN

In honour of a master of the floods. The editor asked me to prepare the 5th edition of this book of the late doctor Johan van Veen, a proposal I gladly accepted

since I greatly value the memory of my dear friend. I therefore hope that I have succeeded more or less in bringing 'Dredge, Drain, Reclaim' in an up to date form as Van Veen would have done himself. Yet there is a great difference. Of course Van Veen could not tell too much about his own pioneer work. He could give his praise to the other masters of the floods and he has done so abundantly. Johan van Veen being a real master of the floods, I got the opportunity to tell something more about his own work. He has always warned against our dikes, most of them being too low to survive a

severe stormflood. What was in such a case to become of the inhabitants living below sea level, what of their dwellings and their cattle? Some of his colleagues called him in a sneering way a newborn Cassandra. But in what a terrible way he was put in the right that horrible night of February 1st, 1953! During almost a quarter of a century Van Veen had studied with his collaborators schemes to strengthen the protection of the isles in the so-called Delta against the sea.

Like that other master of the floods, Cornelis Lely, he needed the help of a stormflood to have the eyes of the authorities opened wide. The famous Delta Plan is now under construction but we should never forget that the Delta Works could not have been started as early as 1954, but for the work by Van Veen and his staff during the past decades. He indeed must be considered to be the father of the Delta Plan.

Doctor Johan van Veen was an engineer every inch of his body and soul. He was interested in the most different fields and saw relations no one else would have thought of. To give just one example, he designed an electronic computer based on the analogy of the tides and alternating electric currents. At the moment a much larger 'analogon' has been constructed to help to carry out the intricate calculations.

On December 9th, 1959, Johan van Veen suddenly died of a heart attack. He had worked too hard and too strenuously. Although he was well aware of it and always warned his friends not to follow his example, he burnt the candle of his life at both ends. In his works he will live on as a real master of the floods.

H. J. STUVEL

CHAPTER I

SPADE WORK

1. Beginnings

Holland – the Netherlands – may be said to be a sand and mud dump left over from the ice age. When the Polar ice-cap, covering the North of Europe to a line roughly from Hull to Amsterdam, began to recede for good, huge amounts of sand, which the melting ice had left on the bottom and the shores of what is now the North Sea, became apparent.

These sands had come from Scandinavia, the Baltic, Poland, Germany, as well as from the Alps, for the place which is now called the Netherlands had been the only outlet for all the melting ice and rain-water of half Europe.

But Nature started work immediately in other ways. Tide and wind combined and built a streamlined fringe of sand, covered with sand dunes, and behind those dunes vast swamps came into being, through which the water of the Rhine, Maas and Scheldt, smothered by muddy shoals, sought their shallow and winding ways towards the sea.

As to size, this delta, now known as the Netherlands, can be compared with the most famous of all deltas – the Nile. The difference between the two was that whereas the Nile delta had to deal with freshets of the river only, the Dutch delta was subjected to additional destructive and constructive forces of the tides.

The turbulent sea with its tides and storm-floods was responsible for everlasting changes, as though Nature could never be content with her handiwork, but must endlessly model and remodel the soft masses of sand she had deposited here.

The sea cut deeply into the swamps, and covered part of them with layers of marine sands and clay. The rivers also brought their material and the inlets became blocked with it. Then the tides broke in at other places, destroying new areas, afterwards to reshape them to form other landscapes. Wide rivers, tidal inlets of great size, sandbanks, dunes, islands, peat bogs and clay marshes came into being, had their time, and vanished.

When the first prehistoric men and women came, they saw this semi-submerged land and did not dare to live there. They lived further East on higher ground and hunted the mammoth, aurochs and wolves. They may have travelled with their scows through the creeks of the hybridic wilderness, but for more than a hundred centuries they did not venture to build their huts in such unsafe places. The Greeks and other Mediterraneans thought this coast a ghostly one.

It has been pointed out that it may have been the coast where the Gates of Hell were supposed to be. The Argonauts sailed their famous ship Argo – some

11

centuries before Christ – through 'a narrow stream' (the Straits of Dover) into the sea which was 'called Death Sea by the people of the North'. Here was the 'eternal fog, where the sea rushes over the sandbanks, covering them'. – 'The people living there *need not pay their fare* when they are dead; their spirits reach the Acheron in a boat immediately and quite near it are the immovable gates of Hades and the land of Dreams.' Another ancient writer says, however, that the land has no inhabitants. But all writers of that time agree that the Cimbres lived near it – the people who are said to have suffered so much from the 'Cimbrian floods', caused by storms.

Homer supposes Hell to be there also; Odysseus is going to that 'land of fog and clouds where the Cimbres live, and where no bright sun ever shines upon these miserable people'. Even as late as the 6th century after Christ, Procopius of Byzantium writes how the men living on what now is called the Dutch coast *need not pay their taxes*, because they serve as ferrymen for the dead. In the depth of night they hear a voice telling them to rise. They go into the ship and feel it become heavier with many invisible people. A hollow voice calls every soul by his name and former function until a freeboard of no more than a few inches is left. In the early morning they sail, reaching the other shore in a few hours. 'I have heard this tale in Britain very often', says Procopius, 'and the people are sure about it, yet I myself cannot believe it'.[1]

Apparently the early navigators disliked and distrusted the low Dutch coast. Of course, they preferred steep visible cliffs without sandbanks to the treacherous shallow sandbanks protruding into a sea with strong ebb and flood currents. While sailing, having no compass needle, they wanted to keep to the shore in order to count the 23 Frisian islands (Pliny says already that there were 23 of them), which hung like a string of beads on the bosom of the North Sea from the Rhine to Jutland. But they could hardly see these low islands at the great distance necessary, because of the sandbanks. There was fog very often too. What seems to have struck the people was:

1. The lowland coast was an awful one for shipping.
2. The people there were navigating and trading people who paid no taxes to anyone; they might have some uncanny connection with ghosts.

One of the first civilized men who saw and described that Coast of Awe in a curious but comprehensible way was the Greek Pytheas. His famous book did not survive, but as some later classical writers refer to it, we know that in 325 B.C. he sailed to Cornwall to visit the tin mines, and then Northward to find the coast where the amber came from (the Frisian coast). He came back with an unbelievable tale; he said that he had seen what he called the 'Sea Lung',

---

[1] Exactly the same tale is still alive in East Friesland where the ghosts go the opposite way. They are loaded at Nesmersiel and brought to 'White Island' (England). The tale is also told on the coast of Groningen, Holland. How incredibly ancient those tales may be!

somewhere in the North, where ice, water and air mingled, without doubt the end of the earth. And he had seen the sea rise and fall regularly! Even several centuries later Pytheas was severely criticized for this extravagance. How could a sea rise and fall? And had he seen the Lungs of the Sea somewhere? Today, we understand. We are not astonished that the Mediterranean people did not know the tides. As to the sea lung, this description of the Dutch coast seems most apt. Those vast sandy foreshores in whose wide creeks the tide moves in and out with a slow breathing rhythm must have made a deep impression on their imaginative minds. The tree-like systems of tidal gullies and creeks closely resemble bronchial tubes. Pytheas could not have described the Frisian Wadden better.

Not much before this 'Greek Columbus' had discovered the Dutch coast, there had been a tribe which actually dared to settle in that soaking wilderness. According to recent archaeological excavations their first farms date from about 4000 B.C. Had there been over-population in the interior of the country and were they a group of trekking farmers? It is presumed that they came from South Sweden or thereabouts; were they a beaten people, chased from their homesteads and obliged to seek refuge in such an impossible country where sudden winds might make the sea rush in to a height of many feet above the land?[1] Or were they merely attracted by the innate fertility of the sea-marshes? We do not know, but excavations show that they built fine large farmhouses – up to 75 feet long – closely resembling those of the existing generation.

Those marsh people soon came to be known as *Frisians*. Their country was the low district along the North Sea, more or less from the present Dutch-Belgian frontier up to the Weser, a stretch of 350 miles. Instead of the name Frisian we may use the name Dutch, or Marsh Dutch and not be far from the mark. The name 'Holland'[2] and 'Dutch'[3] are of later date, however.

The southern part of the poor Frisian country – south of Hoorn – was worse than anywhere else; it was almost totally uninhabitable. In the northern part, where the sea had brought a layer of clay, were the best places, and here the Frisians had their centre. In order to keep their heads above the waters, they made artificial heights. These heights still exist. Much later, the southern, wild, boggy, uninhabitable part became the centre of the country, but this was after

[1] Owing to the shallowness of the North Sea storms may raise its level about 10 feet on the Dutch coast The tide moves then on that higher level.

[2] 'Holland' was originally the country near Leyden.

[3] 'Dutch' is the English pronunciation of 'Deet'. The poet Melis Stoke (13th century) writes his poems, as he says:

> For all the common Deet
> Living between the Elbe and the Seine.

(This is a coastal stretch of 620 miles.)

the year 1200, when man began to learn how to become master even over such inhospitable soil.

The old centre in the North has continued to be called Friesland until this day. The new centre in the south got the name 'Holland'. Still further south the moors were destroyed almost totally by the sea, and this coastal stretch was called appropriately 'Zeeland'. 'Holland' and 'Zeeland' are the provinces at the mouths of the Rhine. Maas and Scheldt and these rivers were the main source of their present significance.

The earliest written records about the Frisians (or Coastal Dutch) describe them as water-men and mud-workers. The Romans found in the North of the country the artificial hillocks upon which the inhabitants, already called 'Frisii', made a living. We shall follow their history, because written records are available about the early reclamation works they made. One and the same race, now called the Dutch, took, held and made the low country.

Pliny, who saw these mound-dwelling tribes in the year 47 A.D. described them as a poor people. He apparently exaggerated when he wrote that they had no cattle at all. Or did he see some much-exposed mounds near the outer shores where the sea had swallowed every bit of marshland? At stormtide, Pliny said, the Frisians resembled groups of miserable shipwrecked sailors, marooned on the top of their self-made mounds in the midst of a waste of water. It was impossible to say whether the country belonged to the land or to the sea. – 'They try to warm their frozen bowels by burning mud, dug with their hands out of the earth and dried to some extent in the wind more than in the sun, which one hardly ever sees.'

No doubt the mud Pliny refers to was the peat which was found in the 'wolds', or swamps, some distance south of the clay marshes, where the artificial mounds had been made.

There were no trees in those sea marshes, as the soil was impregnated with salt. The beams used for building their farms had to be carried from afar. Their very ancient laws speak of 'the North-haldne beams', perhaps the straight fir trees taken from Scandinavian forests.

Of course there were no crops of cereals, except a scanty harvest from the gardens on their mounds. Yet the people must have been a hardy race of fishermen, skippers, cattle farmers and mud carriers.

In all they built 1260 of these mounds in the northeastern part of the Netherlands, an area of a mere 60 × 12 miles. Further East there are more of them in East Friesland. The areas of the mounds themselves very from 5 to 40 acres; they rise sometimes to a height of 30 feet above normal sea level. The contents of a single mound may be up to a million cubic yards.

We can imagine the population of a village bearing willow baskets or hand barrows in a long procession, carrying the clay from the marshes into their

14

villages, raising them gradually, throughout 12 centuries, to keep their families, cattle and farms above the level of the highest floods. These floods seem to have risen higher and higher in the course of a thousand years. There is mud in their baskets, into mud their trudging feet sink, and another kind of mud is used to dry them in the evening. Mud was their fate, since they chose to live permanently on the open marshland near the sea.

Though the sea marshes bore short marine grasses on which their cattle could feed and from which some hay could be made in summer, they were bleak and dangerous in winter, being intersected by innumerable tidal creeks. In this region of water and mud where no human beings before them had ever ventured to live the start must have been risky. Towards the sea were the *Wadden*[1], those sandflats covered twice daily by the flood, where shells and fish were obtained. Landwards was the *Wapelinge*[2], the low swamp or wilderness where no man could pass, except along some swamp trail or in a boat along some tiny creek. Therefore the marsh people were well isolated. In all there were, according to old manuscripts, only seven roads to the interior, four by water (Rhine, Ems, Weser, Elbe) and three by land.

They built their mounds on the shores of the creeks in which the tide ebbed and flowed. In their scows they went (in their language in which the roots of so many English words can be found): 'uth mitha ebbe, up mitha flood' – out with the ebb, up with the flood. The tide bore them towards the peat regions, or perhaps to the woods still farther inland and then brought them back. Or they went out with the ebb in the morning towards the sea, where they gathered their food, and returned in the evening with the incoming tide.

They had no dictator or king to govern them, but formed a kind of natural democracy. The only laws were those of stern necessity and the revered EWA[3], which was *the Law of Eternal Right*, too sacred to be written down, but carried from mouth to mouth in sentences of such form as to be easily remembered. The first letter rhyme was used for this. The law contained such severe ancient sayings as: 'Murder must be cooled with murder', or the equally fierce: 'A murderer must be backbroken, a thief must be beheaded'.

On the other hand we meet with some gentle laws, for instance how a girl

---

[1] Wadden = tidal sands – compare 'to wade'.

[2] Wapel – old Frisian for water. The word Wapelinge is not used any more.

[3] EWA – compare Dutch 'eeuwig' = eternal, for ever.

There were two kinds of law, the Divine and the human. Here is an ancient definition of them: 'Tell me something, I pray about *Divine Right?* – It is the EWA, which teaches a man's mind to judge itself and to conquer what is wrong, to help innocence and to ban cruelty.'

'And what is *Human Right?*' 'The King's commands and the people's customs which are useful and honest.'

'Against the EWA there can be no command which can break it.'

'The EWA is Divine, human right is human; the first is innate in thee and the other shalt thou learn; the first is natural, the latter man-made.'

may choose her husband. Or we read in poetical language about a child when it is carried 'northwards over the sea or southwards into the mountains'. Or 'if the heta (hot) hunger oer the land fare' and the fatherless child may die from hunger, or if the poor child is naked or houseless in the murky night of a 'needcold' winter, when every man flees into his house and even the wild creatures seek their hollow trees of the 'hli' of the mountains. Then the mother of that child must 'setta and sella' the farm and must buy 'ku and korn', to save the child's life. This is her 'pli and plight' – her plea and plight.

The Southern people seem to have been much impressed by the fact that the marsh people of the North need not pay taxes. In our own manuscripts too this freedom from paying taxes is stressed; the reason given is that we had 'to fight the sea as well as the Wild Viking'. – Perhaps the Marsh Dutch of the early days were too poor to pay taxes, or perhaps they defended their drenched isolated country too well against invaders.

They opposed taxes with vigour; when one of the first Roman generals wanted taxes in the form of ox-hides, and later the larger hides of the wild aurochs (a pretence to make the inhabitants sell their women and children), the 'Frisii' rose indignantly and threw the Romans out of their contry – for ever. The Rhine became the Roman frontier.

Much later when the cruel Duke of Alva was sent to quell their revolt and ordered them to pay tithes to the King of Spain, the unflinching revolt lasted 80 years. Remembering the ancient sagas, that sour man Alva called us 'the next-neighbours to hell'. When some Duke asked the North Frisians to pay taxes, and none were given, the Duke himself went to collect them. It is reported that the first farmer he visited was eating his 'brij' (porridge), and that this farmer, after having heard the message, in his great anger took the Duke by the scruff of his neck and pressed his face into the 'brij' intil he was suffocated to death before he could call his escort.

The stubborn opposition to paying tax was not always successful. Clinging to the Right as they conceived it, a tragedy occurred to the Frisian tribe of the Stedingers when they dared to refuse church taxes in 1234. The Abbot Emo, a

---

GEOLOCIGAL MAP OF THE NETHERLANDS (SCHEMATIZED)

The lowest part of the Netherlands (red and violet) was originally uninhabited. The Frisians started about 400 B.C. to build dwelling mounds (black dots) on the highest spots of the sea marshes in the north. The western part remained practically uninhabited, being one vast morass behind a barren dune belt, but it was called Friesland as far as Flanders in the southwest. The peat swamps (violet) were not inhabited and cultivated till 1000–1400. Here a new cultural centre arose with the towns of Amsterdam, Rotterdam, The Hague, Haarlem, etc., called Holland. The southwestern regions are fresh marine deposits on the original peat swamp. This part of the land was reclaimed mainly after 1500. The eastern part of the Netherlands is sandy soil, originally covered with woods and heather.

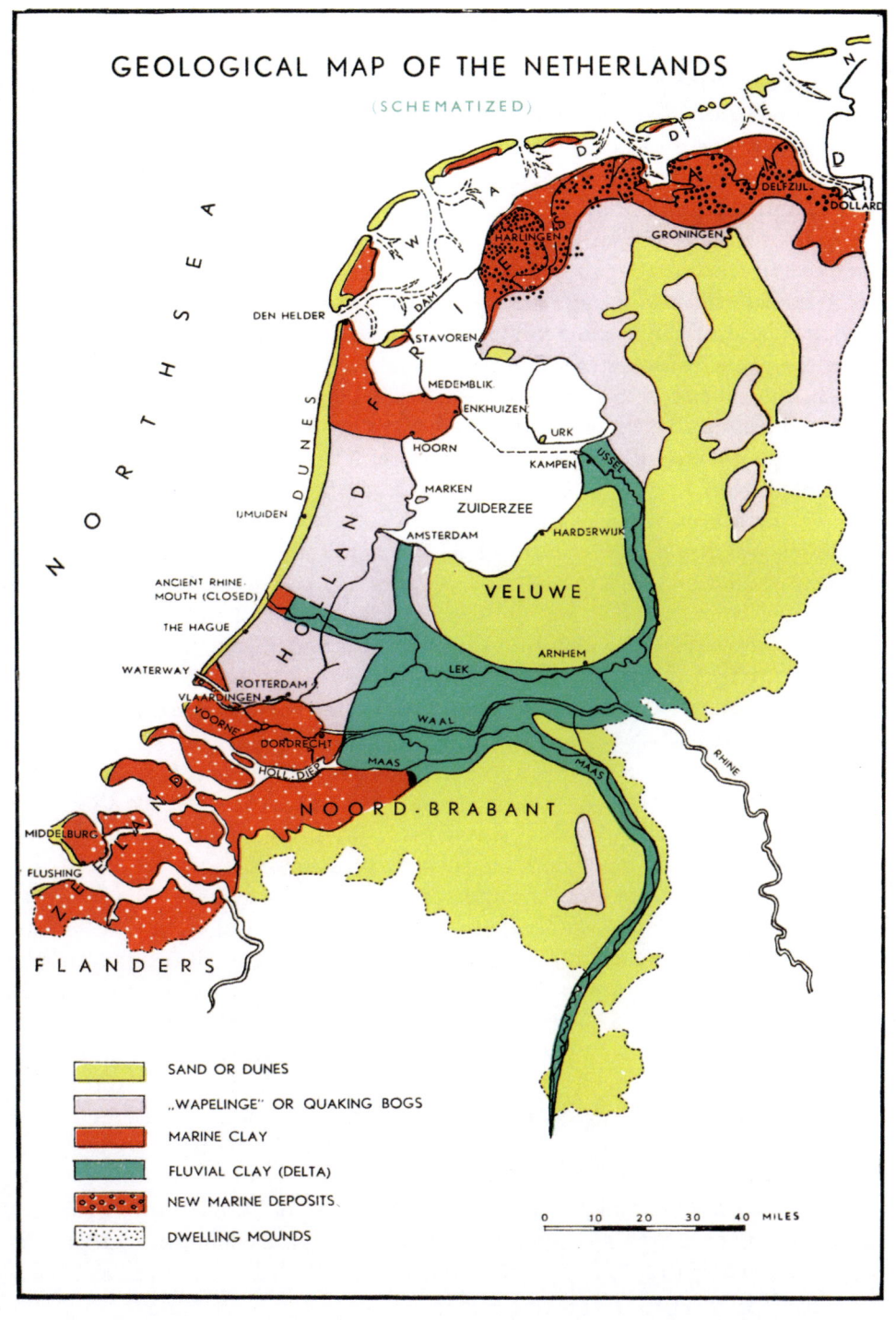

GEOLOGICAL MAP OF THE NETHERLANDS

(SCHEMATIZED)

Legend:

SAND OR DUNES
„WAPELINGE" OR QUAKING BOGS
MARINE CLAY
FLUVIAL CLAY (DELTA)
NEW MARINE DEPOSITS
DWELLING MOUNDS

0   10   20   30   40   MILES

Map labels:

NORTH SEA

DEN HELDER · STAVOREN · MEDEMBLIK · ENKHUIZEN · URK · HOORN · KAMPEN · MARKEN · HARDERWIJK · ZUIDERZEE · IJMUIDEN · AMSTERDAM

DUNES · HOLLAND

ANCIENT RHINE MOUTH (CLOSED) · THE HAGUE

VELUWE · ARNHEM

WATERWAY · ROTTERDAM · VLAARDINGEN · VOORNE · DORDRECHT · HOLL. DIEP · LEK · WAAL · MAAS · RHINE

MIDDELBURG · FLUSHING · ZEELAND

NOORD-BRABANT · MAAS

FLANDERS

HARLINGEN · GRONINGEN · DELFZIJL · DOLLARD

FRIESLAND · DRENTE · ISSEL

Frisian himself, living near Groningen, relates this in his diary in prudent words. He dares to write: 'The Preachers flew in packed multitudes' over the whole of northwestern Europe, 'they flew around like clouds', preaching a crusade against the 'disobedient Stedingers' under the motto 'Disobedience is idolatry'. Those who would enlist to kill the Stedingers obtained general indulgences. The loot attracted robber barons and even counts from afar, together with much rabble, and the population (11,000 warriors) was murdered almost to a man; only a few could save their lives by fleeing into the Wapelinge. Emo says: 'This was the third crusade against disobedience and idolatry, the first was against the Saracens, the second against the heretic Albigensis (1208–1229) and the third was against the Stedinger Frisians'. Horrible times.

Yet the other Frisian tribes managed to live on without paying taxes. Abbot Menco, the successor of Emo who continued his diary, writes in 1267: 'It is quite understandable that because the Frisians are the only people in Christendom who do not pay tithes and firstlings (Deut. XII, 6), the sea takes away what is withheld from God'. – Horrible dictum. But Menco, as well as Emo, helped to make the dikes stronger.

The sea was the most dangerous foe. It wanted and took more tax than all other foes together; land as well as lives it asked – thousands of acres, many thousands of men, women and children and hundreds of thousands of cattle.

When 'the waters prevailed upon the earth, it was the flesh that died, both fowl and cattle, beast and every creeping thing that creepeth upon the earth and every man' – at least those who could not reach the artificial heights quickly enough.

The Coastal Dutch have now lived 24 centuries in their marshes and of these the first 20 or 21 were spent in peril. It was not until 1600 or 1700 that some reasonable security from flooding was achieved. During these long treacherous centuries the artificial mounds made their survival possible.

The villages still stand upon the ancient mounds. The highest spot is reserved for the church, which towers above the houses. The cold churches and houses are red from top to bottom, because the stones and tiles were made of the same clay as the mounds themselves, but now baked into a brilliant red. Where there was much chalk in the clay the colour is yellow – which is also comely. What else was there to work with but clay? We took our first steps in clay, made hills from it, made our pottery from it, made pits in it to gather rainwater, made sea walls from it, and later used it for making stone houses and hard roads. The first monuments which our forebears made are in clay – the hundreds of artificial hills to be found in our Northern marshes.

It was a work which might be compared with the building of the pyramids. The pyramid of Cheops has a content of 3,500,000 cubic yards, that of Chephren 3,000,000 and that of Mycenium 400,000 cubic yards. The amount of clay carried into the mounds of the northeastern part of the Netherlands can be estimated at 100,000,000 cubic yards.

In Egypt it was a great and very powerful nation which built the pyramids throughout a series of dynasties. The aim was to glorify the Pharaohs. With us it was a struggling people, very small in number and often decimated, patiently lifting their race above the dangers of the sea, creating large monuments, not in stone, but in the native clay.

2. *The 'Golden Hoop'*

After the hillocks had been made there was no respite. A heroic time approached. It started about 800 when Charlemagne's Christian warriors conquered the heathen Saxons and Frisians.

Then Charles the Emperor ordered the EWA to be put down on parchment. But this the stubborn inhabitants of the sea marshes refused to do, the 'ETERNAL' being far too holy to be written down for every outlandish foe to read and corrupt. The twelve Wise Men who dared to withstand Charlemagne's command for six successive days were in a bitter plight at the end of those days, when they had the choice of the following three penalties: to be put to death, to be made slaves, or to be set adrift at sea in a boat without oars or rudder.

They chose the last. Awaiting death far from shore, one of the Wise Men remembered in his extremity that he had heard a tale about some new Man, helping people in distress on sea, and lo! a thirteenth man, not different from themselves, was seen sitting in the stern of the boat, steering the ship towards the shore. After this He taught them how to set down the law in writing.

Such is the saga which has come down to our days, and we know that what is called the Lex Frisionum was put on paper in the year 802[1]. It might be asked, however, if it contains all the EWA. There is much of the 'human law' in it. The higher Law which must have contained some Religious Rules of Right, is absent.

This ancient saga of the twelve stubborn Wise Men reveals the devotion to the EWA and respect for the Highest. The scruples to keep the holiest undefiled and the clear definitions of 'Eternal Right' and 'human right' seem to be the corner stones on which the civilization of the Low Countries has been built. They mean the acceptance of the principle of Right above Might, of Democracy above Despotism, of Freedom above Slavery. For the EWA (or ETERNAL) taught a man's mind to judge itself and also that human right was but the King's commands and the people's customs.

---

[1] It contains such queer rules as: the killing of a freeman 53 shillings, killing of a hunting dog which can kill a wolf 8 shillings, killing of a pet dog 12 shillings, cutting off an ear 36 shillings, cutting off the nose 78 shillings, touching the breast of a woman 4 shillings, etc.

It is typical that the EWA had to be 'found', it was not decreed by some Ruler. Like students in mathematics the Wise Men or Asegas set out to *find* the right solution of a problem. At the annual gatherings, called 'Thing', everybody was supposed to help in finding the EWA in some particular case: 'He who knows better must say so'.

This fundamental idea about right and wrong, being 'innate in thee', is still strongly felt. It has led to dour inflexibility of character – to a kind of super-individualism and an ineradicable obstinacy.

In this Lex Frisionum of 802 there is not yet any mention of *seawalls*, but the first attempts at dike building must have been made shortly afterwards. Frisian manuscripts still extant, dating from the early Middle Ages, deal chiefly with the following three points:

First, the right of the people to freedom, all of them, 'the bern and the un-bern'. Secondly, the 'wild Norsemen' whose invasions took place roughly from 800 to 1000, and thirdly: the Zeeburgh or Seawall.

This novel means of defence against the sea by means of a continuous clay wall was called a Burgh, or stronghold. The people were apparently very proud of this seaburgh, because they described it in poetical language as 'the Golden Hop', the Golden Hoop. – 'This is also the Right of the Land to make and maintain a Golden Hoop that lies all around our country where the salt sea swells both by day and by night.'

After a long and certainly none-too-quiet life on the isolated mounds of 14 centuries, three factors made themselves felt almost simultaneously about 800. They were:

1. Christendom: the pioneers were British preachers, who could be understood easily along the Northern shores of the Low Countries;
2. the Viking invasions from the North, and
3. the beginning of the fight against the water.

Formerly the terrible evils of the sea, the storm floods and the more terrible marine erosion, had to be endured, but now the fight began to throw the sea out of the country; a fight not yet ended and a fight for to be or not to be, which will last many thousands of years. The cause of the building of the seawall may have been overpopulation, the great cause of all development of any country. The dike building mainly started after the Norman raids slackened.

The ancient Laws which have come down to us contain the following oath: 'We shall defend our land with three weapons: with the spade, with the hand-barrow and with the fork. Also shall we defend our land with the spear and sword and with the brown shield against every unjust lordship. Thus shall we defend and keep our land from end to end, so help us God.'

The spade, the hand barrow and the fork were the instruments used for

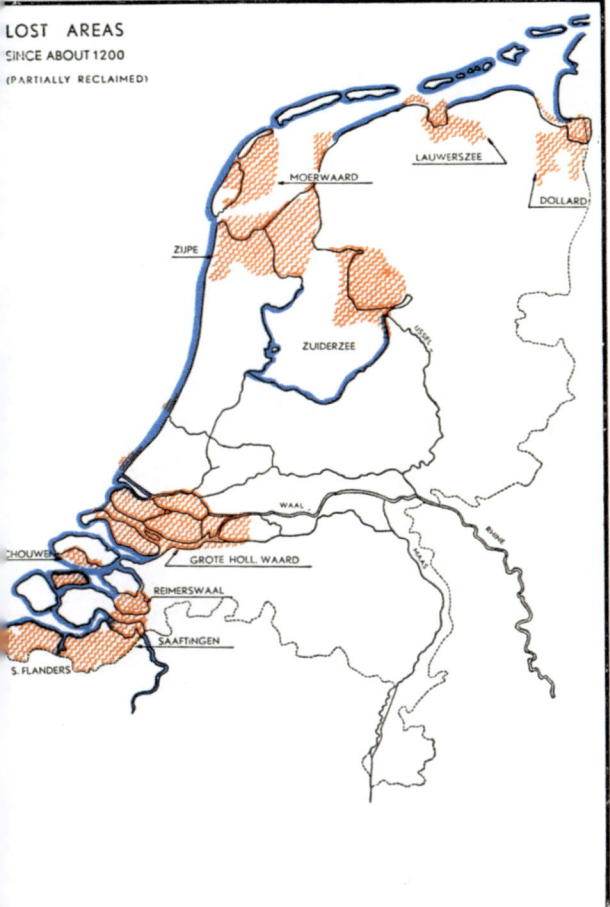

LOST AREAS
SINCE ABOUT 1200
(PARTIALLY RECLAIMED)

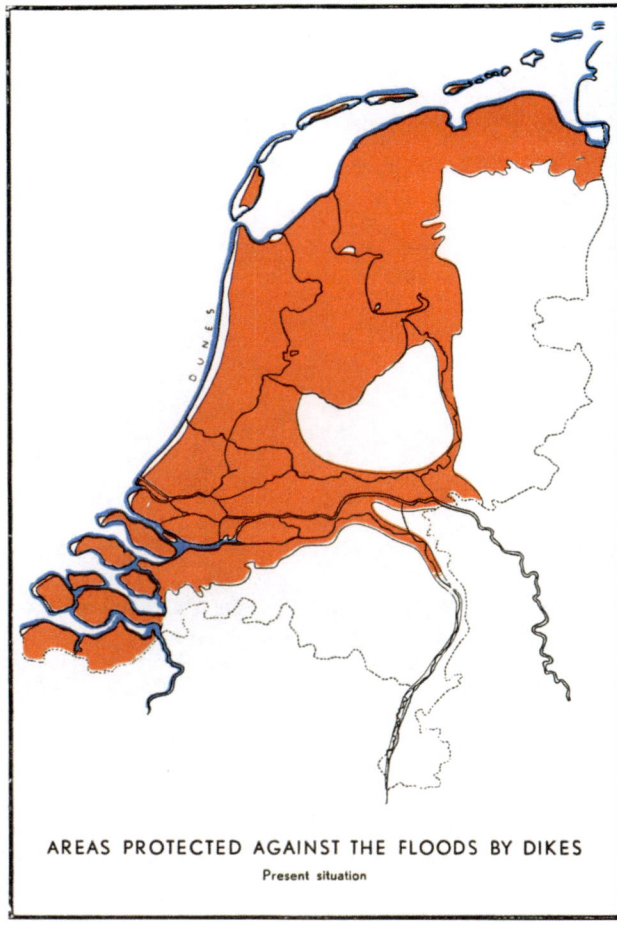

AREAS PROTECTED AGAINST THE FLOODS BY DIKES
Present situation

The low moor country became an easy prey to the sea. Many of the people living in it were drowned, because the dikes always broke suddenly. The dikes of the rest of the low country also broke often, but those breaches could always be repaired.

diking, the fork presumably for the grass turfs which were used to heighten the dikes and make them stronger.

How dangerous a foe the sea was is clear when we read that the Frisians were not required to go in the army farther than their own frontiers, i.e. the Weser in the northeast and Flanders in the southwest. The Emperor had agreed to this because the people in the marshes had to defend their country both against the sea and against the 'wild Norsemen', the manuscripts tell us. It is also apparent from the fact that a man who had to work at the improvement of his section of the dike was exempted from going to the Tree. This Tree was the Holy Tree, known as Upstall-Tree, under whose branches the annual Thing of the 'Seven Frisian Sea-lands' was held and where attendance was compulsory

ANCIENT TYPE OF DUTCH MARSHLANDS

The picture shows a road along wich the houses are built. The swamps have been split up into neat parcels, separated by wide and deep ditches. The dredgings from those ditches have been deposited on the islands, in order to heighten them and make them more fertile. The fields can be reached with scows only, and in the central pool a small harbour has been made where the scows bring cows, hay or vegetables. Originally such land was called 'Wapelinge' or 'Waterland'. The ditches may have been made in the early Middle Ages, because the pattern of the fields is rather regular. The village shown here is called Brook. *(Photo K.L.M.)*

for every man, except for the 'needcase' that 'wind and water had turned against him so that he should be at the dike'.

There is another ancient oath in which the fight against the sea is mentioned. It is akin to the former; the translation of the original, though the two languages are akin, loses in beauty:

'With five weapons shall we keep our land, with sword and with shield, with spade and with fork and with the spear, out with the ebb, up with the flood, to fight day and night against the North-king (the sea) and against the wild Viking, that all Frisians may be free, the born and the unborn, so long as the wind from the clouds shall blow and the world shall stand.'

It reminds us of Nehemiah when he built his wall round Jerusalem: 'And it came to pass from that time forth that half of my servants wrought in the work and the other half of them held both the spears and the bows and the haber-

22

ANOTHER PICTURE OF DUTCH MARSHLANDS.

As in the preceding photograph, there is a road along which the houses are built. The fields are surrounded by irregular small lakes and ditches, indicating an early settlement. The road, consolidated for ages with sand (brought from afar) is 'floating' on the spongy material underneath. Some of such half land – half water regions are now being reclaimed by pumping more water out of them. The whole country may then shrink some yards.   *(Photo K.L.M.)*

geons. They which builded on the wall and they that bare burdens with those that laded, everyone with one of his hands wrought in the work and with the other hand held a weapon. For the builders everyone had his swords girded by his side and so builded. So built we the wall, for the people had a mind to work.'

This new era after 800 was one of unification. It meant education in co-operation. The stubborn free farmers were compelled to stand together, if they were to keep their land. Previously every dwelling mound was self-contained and lived a life of its own. There must have been fights between one mound and the other. Now they had to combine, if they wanted to remain free and keep their 'Golden Hoop which lies all around the country'.

'The work is 'great and large', said Nehemiah, 'and we are separated upon the wall, one far from the other.' By reasoning with the people he caused them to collaborate, and his success was great. Whether there was a Nehemiah among our dike-builders is not known. The earliest Christians must have worked in the

23

The 'Golden Hoop' sheltering a small village. In the foreground a groyne, made to keep the currents from the dike.

right way. In any case there was no Emperor or Pharaoh to order or instruct them in making the seawall. It was the second monument in clay which they erected, larger in content than the pyramids of Egypt, and it was the EWA, AWA, or the Law of Eternal Right which governed this mighty work.

The Awa-sayers, or A-segas *found* the law of Right for diking and they became what were later called the dike masters. They stimualted the sense of orderliness, as may be seen from the following lesson which can be read in one of the old manuscripts:

'Asega', asks a farmer, 'what have we to do in this new year?' 'You have to keep the Law of Peace towards your neighbours and your family', says the wise Asega, 'and to do everything which is seemly and right.'

'What have we more to do?' asks the farmer again.

And now the Asega answers:

'Cleanse the sluices, clear the ditches, repair the highroads and home roads, heighten and strengthen the sea-burghs and dams, and make drainings underneath the roads that the water may pass; in springtime work at the sea-walls, in summertime cleanse the drainings from vegetation and throw this on the sides, and during midsummer work the whole long day in the fields.'

24

In times of flood the cattle were brought up into the roof stable, the door of which is shown here. A special bridge was used for bringing the cows into their attic.

Flood-stones, indicating the levels to which floods rose, in buildings standing at the shores of tidal inlets, give us an idea of how these buildings settle. Dikes settle more because heavier.

Such is still the law of the Low Countries! The dike masters repeat the same orders for cleansing the ditches, repairing the dikes and roads and keeping the sluices drawing with ebb and stemming with flood every year afresh, and these orders are being carried out, not so much because of the dike masters, but because of necessity. There may be some sluggards who cannot or who will not do all that cleaning and repairing, but the work *must* be done, because it is the EWA, the Law for the 'berna and the unberna, so long as the wind from the clouds blows and the world stands'.

The world still stands, the wind is still blowing, the clouds are sailing along the skies as they did long ago, and our race is still there to obey the EWA of necessary cleanness and fitness of all things applying to dikes, dams and drainage. Likewise we have learned to some extent to obey the EWA in matters regarding unity of purpose. This last matter, because of our individuality, we

25

found and still find difficult. But if we want to keep our country trim, neat and fertile, we shall have to stand together, willy nilly.

3. *Perseverance*

A few years before the second world war a British diplomat went to Germany with his yacht by way of our coastal waters. He wrote about this trip. Standing upon the high seawalls protecting the country from being flooded and seeing the amount of energy which had been expended to make the land, to keep the sea out and to drain the low soil and cultivate it, he said it was obvious to him that all the fine reclaimed land could never repay the gigantic work which had been put into it. But, he added, the toil has not been in vain, for has not this coast been one of the cradles of democratic freedom and of governing people by Right and not by Might?

We ourselves think that the labour of our ancestors has been worth while, even from a material standpoint. The fields have repaid and are repaying abundantly. As regards the necessity of co-operation, it may have helped to foster the idea of government by Right or its approximate concept, but we cannot pride ourselves on having had an abundance of brotherly feelings. Necessity ruled us and it was a hard taskmaster. Damming roaring streams by means of willow mattresses and clay mixed with straw was a fine sport and comparatively easy for us, as the earliest written records reveal (the first sluices are mentioned as early as 806), but it must also be mentioned that there have been many feuds which have often prevented the execution of positive work.

There were many lapses by the wayside. At such times the strength of our individuality became our weakness.

We will not dwell here on the shameful feuds. Whoever wishes to know more about them may read the old chronicles, such as that of Emo and Menco, which covers the thirteenth century. Crusades, wars, diking and floods – always many floods – permeate that age, and it would seem that in those medieval times the sea still proved to have the upper hand. This was due partly to our insufficient technical skill and partly to lack of co-operation. For a single night, Dec. 14th 1287, the officials and priests estimated that 50.000 people had been drowned in the coastal district between Stavoren and the Ems. This is a large number considering that this was the area where so many dwelling mounds could be used as places of refuge.

According to a recent Soil Charting Survey the marine erosion of the marshes between the years 300 and 800 was such that life became almost impossible. The country which had always lain under the sea's highest floods was on its way to total destruction when man was confronted with the choice to quit or stick and fight. A formidable defensive and offensive struggle for life was started about which we know hardly anything, but the 'Luctor et Emergo' must have been more severe than present-day Dutchmen, or perhaps any other living race, can imagine.

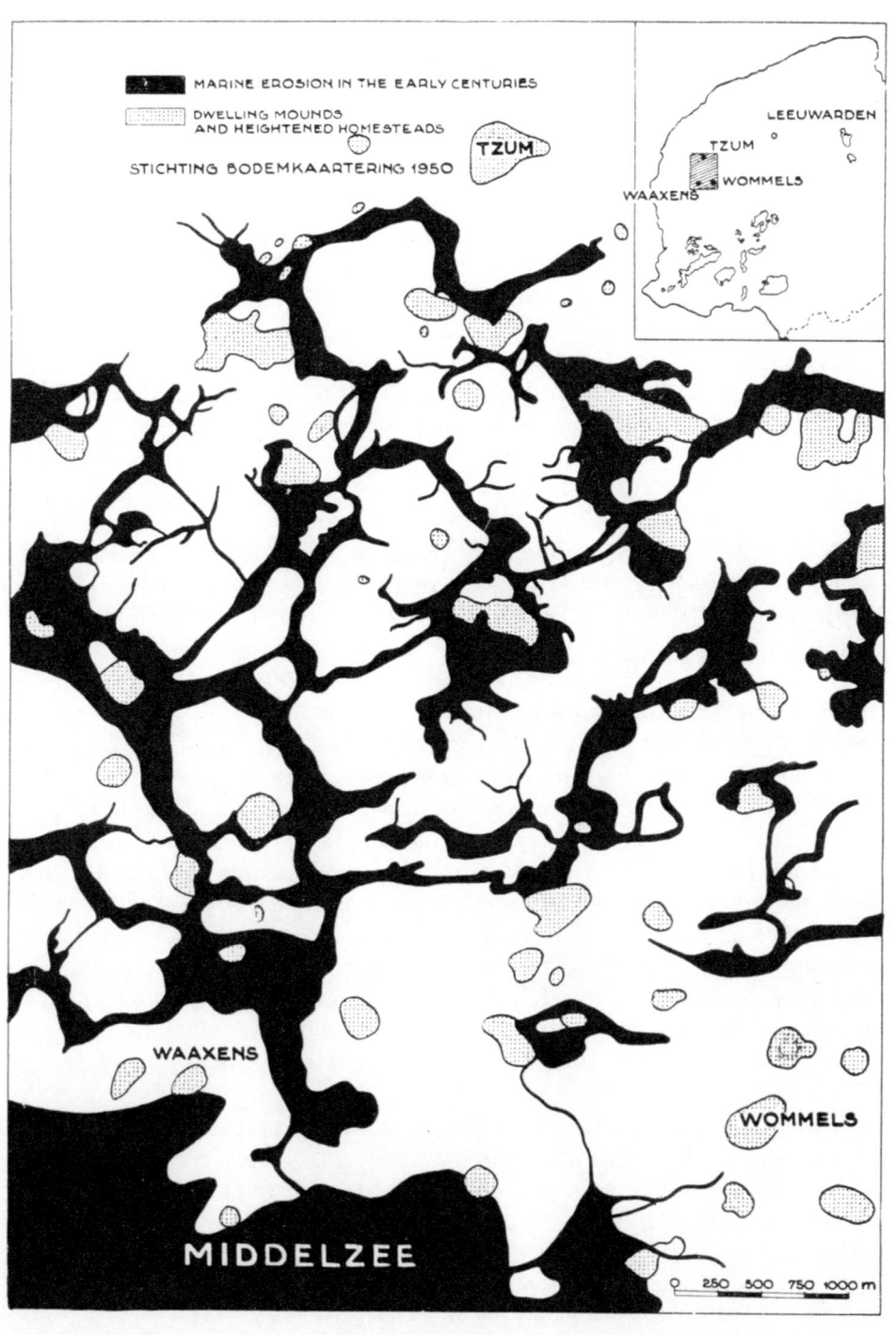

MARINE EROSION IN THE EARLY CENTURIES

DWELLING MOUNDS AND HEIGHTENED HOMESTEADS

STICHTING BODEMKAARTERING 1950

TZUM

LEEUWARDEN

TZUM

WAAXENS WOMMELS

WAAXENS

WOMMELS

MIDDELZEE

0 250 500 750 1000 m

There were a few successive years of rest, when *'people ventured to live on the land again'*, repairing the broken dikes, building new farms and growing new crops of beans and barley, then suddenly the Nor'wester would come once more and submerge their labour, their cattle and their children or themselves. A large percentage of the people must have been drowned in the course of the first stages of the building of the 'Golden Hoop'. This 'hoop' or sea-burgh was not nearly strong enough. It collapsed repeatedly before 1500. Even as late as 1825, 1894 and 1916 there were serious breaks. The early attempts to make and hold the country were far more strenuous than might have been expected in the beginning. After the better drainage of the low land, made possible after the building of the seawalls, it had shrunk considerably, so that when a dike broke, the sea remained inside the submerged land. Those breaches, where the tide found low country inside the dike, were difficult to close, because the tide ran in and out through such gaps to fill and empty the tidal basin. There was also the natural rise in the heights of the floods when they were hemmed in by the new dikes and dams. The inhabitants could not foretell the height of these rises as we can now do by the use of intricate mathematical formulae.

The offensive war which man in his relative ignorance of the power of the sea had started, was answered by a counteroffensive. The more we raised the dikes and the better we drained the country, the higher and more destructive became the floods. We now know the reason, but our ancestors of a thousands years ago had not this exact knowledge, nor our accumulated experience.

Very severe laws were enforced to win the battle. No feuds were allowed once the dikes needed repairs. This so-called *dike peace* could not be broken. Anyone breaking it by fighting was sentenced to death. He who could not help at the dikes had to leave the country; he who would not help was put to death. In some parts any man refusing to do his share could be buried alive in the breach with a pole stuck through his body. The people who lived farther inland had to come and work at the dike as well as those who dwelled near them. 'Dike or depart' was the old saying. Any man who was unable to repair the breach in his section of the dike had to put his spade in the dike and leave it there. This was the sign by which he gave his farm to any man who pulled the spade out of the dike – that means to say: who felt powerful enough to close the breach. This was the *Law of the Spade*, a hard law indeed.

The struggle for existence was hard. Many farmers, having lost their property, had to find a living elsewhere in a foreign country. Many sons of some proud and ancient families had to become labourers. One night could rob them of everything, even their own lives. Plague and famine remained for those who managed to save their lives, for a flood was always followed by death and epidemics.

There is an ancient Latin saying: 'Frisia non cantat' – Friesland, that is, the

28

Polderboys repairing a dike.

low part of the Netherlands, does not sing. Foreigners noticed early that this was no place for mirth and song, but for work and strife. The art of singing seemed to have been forgotten.

No one, not living in the Low Countries and not having studied their history, can understand the struggle these placid lands have had. The land has a quiet, fertile appearance. There are no signs of the fierce contests of the past. But this peaceful country, where fine cows graze contentedly, has been submerged and recaptured perhaps half a dozen times. Time and again such a polder may have been *lost* for half a century of for even more than that. Some historian says: 'Delve into ancient records and we encounter phantom villages. We see the houses and the church, the names of the farms and of those who live in them. We see the pastor in his pulpit. – Then suddenly, all vanishes. Even the location of these villages can no longer be found, nor the beds of the big rivers that flowed past them. All is buried and gone.'

But nothing is lost of which a trace shall not be found! The soil gives up its secrets in time. Whole lifetimes are spent in exploring those secrets of the soil.

Viewed retrospectively, there were severe losses of land in the one hand and several gains on the other. Nature gave new land in some places and took away old land in others. The Zuider Zee was formed about 1300 and the Great Hollandse Waard near Dordrecht was destroyed in 1421. The wide Dollard

29

basin was formed in the North about the same time, and later came the loss of the extensive lands of Saaftinge and Reimerswaal in the southwest, etc. Where fertile land, bearing many villages, had reached far beyond the eye, salt water reigned after one single night of terror. In the night of 18th to 19th November 1421 no fewer than 65 villages were submerged in one single polder, together with about 10,000 men, women and children drowned – a disaster which should be viewed against the population of that time. In 1514 the whole population of the western part of Holland was estimated at only 300,000 and of England and Scotland at 3,000,000.

In many of these submerged places the water still holds sway. Hardly a single ruin of a church or castle remains as a monument to a once-thriving population. All is covered beneath a thick layer of sand and silt. Of that flood of 1421 the only reminder is a church bell, saved from one of the submerged churches shortly after the disaster. From such times date some ballads, e.g. the one with the wailing refrain:

Och, the Ja is so deep!
Och, the Ja is so deep!

The Jade, or Ja, is now a wide inlet in East Friesland. It was formed about 1650.

Not all the submerged land remained under water. In some places Nature was generous and healed the wounds which it had inflicted. It took and gave without asking whether we liked it or not and this is the best proof that our strength was inadequate. The people may have felt from the start that they would get the upper hand in the end, and especially in summertime they may have believed that they had stemmed the danger to some extent, but in reality they were still too ignorant and lacked strength.

The great debacle of 1421 was possible because we had made a large polder by damming the wide river Maas near its mouth in 1213. We had dammed this river at both ends of a stretch of about 30 miles and had led its upper part into the Rhine. This had been too daring and we might well ask how we ever managed to carry out such a tremendous task at such an early date and how,

---

The baby with the cat in the floating cradle seems to be more than a legend. This genealogical painting, apparently a reproduction of a still older painting, was found with one of the descendants of the dredging, salving and towing families of the Alblasserwaard (Kinderdijk). The genealogy (square = masculine, oval = feminine) is fairly complete for the period 1421 to 1692, and famous names like Elsevier, the printers, are included. The dates of the men and women indicated are mostly known. The baby was apparently a girl, was christened Beatrix, was married to Jacob Roerom and had three children. The last couple mentioned are relatives: J. de Custer (1687) married M. E. de Haas (1692). – Beatrix, an uncommon name among the Dutch, is called the 'stam-moeder', or 'mother of the race'. The mansion on the river where legend says she came ashore is 'Het Huis te Kinderdijk', still existent, though renovated. The town to the right is Dordrecht.

for 207 years, we ever managed to keep this land from total destruction. But this destruction came in due time – sudden and irreparable.

The battle was offensive as well as defensive; we were thrown back again and again – overwhelmed – and yet the struggle went on. One of our poets, after asking: 'Wie is er beter dan een boer?' (Is there any man better than a farmer?) shows the endless toil and perseverance of the much-abused farmer throughout the ages. Whatever happened in history, 'the farmer, he ploughed on'. He ends up by saying that the farmer, after having taken a brief rest beside the plough, heard a voice saying to heaven, earth and sea: 'For the sake of the farmer who ploughs on, let the world continue to exist'. In the same way a voice seems to have said: 'For the sake of the Dutch farmers and dike masters, let Holland continue to exist'.

While the floods destroyed all the vast land south of Rotterdam and Dordrecht and the waves raged against the dike of Alblasserwaard which proved to be too strong for the time being, people standing on that dike saw a cradle coming from the west borne by the storm – a wooden cradle of those days. In it was a cat jumping from side to side, rocking the cradle in time with the waves, thus saving itself and the baby. The dike where the cradle with the cat and the baby ran ashore has since then been called 'Kinderdijk', which means 'the baby's dike'.

This legend goes to explain that those who escaped drowning came ashore at the dike of Alblasserwaard and remained there. Those were the ancestors of the most famous tribe of water-fighters the world has known. There are not many Dutch hydraulic engineering firms whose founder did not originate from Alblasserwaard or from places immediately in front of it. Dordrecht itself had become a small island when the flood of 1421 engulfed all the land up to the very walls of the city. The drowned land became a 'wilderness of bullrushes' (Biesbosch). Reeds, bullrushes and willows have grown there abundantly. It provided the elementary school where the Dutch contractors of hydraulic works developed their ability. The peculiar region and the peculair needs of the Low Countries caused a peculiar race to exist here.

Andries Vierlingh, one of our greatest dike masters, wrote in 1570 or somewhat earlier that there were no better dike workers than those of Alblasserwaard, and the same is true to day. The workers of willow mattresses – 'the polderjongens' – who can strangle wild streams, or who can handle the unwieldy dredging machines with unerring skill and in the cheapest possible way, as well as the contractors themselves, all come from the old dike of Alblasserwaard. Nowadays they are from Kinderdijk, Alblasserdam, Dordrecht, Papendrecht, Zwijndrecht, Sliedrecht and Hardinxveld: in Vierlingh's days they came mainly from Alblasserdam. These people have kept up their unbroken tradition for centuries; they will not break with it in the future. The 'polderjongens' – those

32

KINDERDIJK (BABY'S DIKE)

On the right the *old* Alblasserwaard country; irregular ditches (pre-medieval) and some regular (medieval). Four deep holes, filled with water, indicate so many breaches of the protecting dike. This dike, or seawall, called Kinderdijk, follows the right bank of the river Noord which is 225 yards wide; the dike bears a road. On the left *new* country, obtained by making inundated 'lost' land silt up anew; the fields are modern, the dikes straight. According to the legend the child in the floating cradle, after which the dike was named, ran ashore here, when the Alblasserwaard dike withstood the storm flood and the rest of the country was lost. This old dike has been the home of the Dutch dredging contractors and 'polderboys' since at least 1500. The woolly areas on the left shore of the river are willows, used for making willow mattresses.   *(Photo Allied Air Force)*.

long-limbed, heavy-handed, slow-speaking workers in long boots, who have travelled over the whole world to do their mud and mattress work, come home after that work is finished and then stand on 'THE' dike until a new call comes. Any sense of frustration at the whims of nature is absent in them; saving land or making new land has been their job for more than five centuries.

I doubt whether there has ever been a conscious sense of hard life amongst our farmers and dike workers. The inundations and hardships are an accepted part of their life, just as the dangers and hardships of the sea are second nature to the sailor. *William the Silent*, the founder of the Dutch Replubic, was of the same character. When fighting the mightiest state in the world, Spain, and when

33

a. Landmaking by planting spartina plants.   b.   So-called living dam, two years old.

everything seemed lost he said: 'It is not necessary to have any success when trying, nor to have hope when you want to make a start'.

4.   Success

William the Silent wrote those words in the days which were later called 'the time of trouble'. He was more than a great democratic war leader and diplomat. He was a maker of new land. His family had an extensive property in one of the southwestern districts of the Netherlands, and the tradition of that family was to increase its property by coaxing the sea to give up more land. All landowners and farmers who had the opportunity did the same for many centuries.

William's dikemaster was *Andries Vierlingh*, already mentioned. This dikemaster wrote a book about making dams, dikes and sluices, and also about creating new land from sandbanks or sandy foreshores. His manuscript, written about 1575 in his old age, remained in some obscure spot to be discovered in recent times, about 30 years ago.

Vierlingh is not the originator of the Dutch way of dealing with water, but he is one of the endless line of dike masters who developed extensive and practical knowledge of how to handle tides, sands, clays and willow boughs. We of the 20th century, who have had the benefit of instruction in modern hydraulics, are astonished that we can learn so much from an ancestor who lived 400 years ago. Our methods of making dams and dikes have not changed much since his time.

We knew before the discovery of Vierlingh's manuscript that great reclamation work had been done in the midst of the persecutions and wars of the 'time of trouble' in which Vierlingh lived. In our imagination we had already seen a dike master of that time in his wide baggy trousers and high pointed hat standing upon a newly-made dike, looking with satisfaction into the new polder. It is a supreme moment when the dike master stands for the first time upon the completed dike. After the long process of making the new land by the cultivation of reeds and rushes; after digging and redigging again and again so many silt trenches; after 'throttling' several tidal creeks and after the turmoil of a thousand dikeworkers making the new dike in a single summer – a silence falls at last

34

c. Field two years old.

d. Result of initiative of a farmer taken in 1926, photographed in 1943.

and this is the time for the dike master to contemplate his work, though for a short spell only.

Looking into the virgin land he feels the satisfaction of the man who has been a creator. A small and secondary creator compared to the Great Creator who 'saw everything that He had made and behold it was very good'. But new soil has been won at last and in his imagination the dike master sees the new inhabitants come and he remembers: 'Go in, possess the land and multiply'. In the centuries to come this land will remain very fertile, new peoples will live and work here for countless generations, and they will bless this land for its fertility. 'Unto eternity' is the hope of every dike master.

Every man who has made a new polder cannot help wishing that this new soil should be a 'joy forever' an endless source of crops. He feels that this will be so and that something worth while, something substantial, has been accomplished. We, who belong to one of the 'generations to come' are bound with strong ties of gratitude to the ancient dike masters. That is why, when Vierlingh's splendid book of 404 pages was discovered it was studied enthusiastically. Vierlingh was found to be a real master of the dikes and waters, a man of great ability and spirit – one of the greatest of his kind. Luckily the greater part of his manuscript has survived. Its ancient picturesque style is a joy to every hydraulic engineer. This remarkable book already shows the special vocabulary of the Dutch diking people in all its present-day richness. In some ways it is even richer.

His advice is simple and sound. The leading thought is:

> *Water will not be compelled by any 'fortse' (force), or it will return that force onto you.*

This is the principle of streamlines. Sudden changes in curves or cross-sections must be avoided. It is the law of action and reaction. And truly, this fundamental law of hydraulics must be thoroughly absorbed by any one who wants to be a master of tidal rivers.

WILLOW MATTRESS

In this picture the mattress is made in situ at low water to cover a sand dam against the scouring of the flood. There is a 'lower grid' – seen here – then two or three layers of willow, and upon those, an 'upper grid' of the same material. The mattresses are weighted with stones.

He explains explicitly how by delicate, simple, but intelligent ways we can gently lead the tides. 'Look at the 'dammekens' (the small groynes), they are but wattled twigs of willow, weighed down with clay sods but what great benefit they bring at low cost! You can do wonders with them'. 'Look at the small bricks', he adds, 'they are but baked clay, but you can build a castle with them'.

This advice of using gradual methods 'at low cost' is often neglected in civil engineering. It takes a great deal of personal interest and study to find out what is the cheapest and best way. For instance, a single large groyne has at times been made to keep the tidal streams far from the shore. Thus 'fortse' was used upon the water, and the water in its fury made holes 30 or even 60 yards deep immediately in front of that groyne. Then it is necessary to throw stones (bought in foreign countries for millions of guilders) into the water in order to save the groyne. But often the eddies behind that obstacle are so strong as to erode the shores which that groyne is supposed to protect.

So Vierlingh is the herald of streamlining, several centuries before modern hydraulical science discovered its uses. In fact, river and shore hydraulics came after aeronautics in using streamlining. Vierlingh does not halt at using stream-

Willow mattress being sunk with stones, an age-old Dutch proceeding to protect the soft shores against erosion.

lining in matters of leading watercourses; he also stresses the fact that we should use the same principle of gradual progression as regards *time* and *money*. Jerky proceedings are bad proceedings. What a show of incompetence it is to use brute force only! Is the making of large expensive dams an art? Anyone who has money enough can do it! The tax payer will not notice it perhaps, but what a poor sight it is to see an engineer working on the principle: 'The Government has to pay, not I.' What a poor thing is mere violence, even if it leads to success. He shows a better way: 'Direct the streams from the shore without vehemence. With subtlety and 'sweetness' you may do much at low cost'. This is the sporting attitude and this attitude should be taken as regards time as well.

'Do something every year, make your work grow steadily. The art is not to make expensive large dams, but to work gently and cleverly trying to obtain great advantage. What is needed is patience and the use of time'.

Apparently Vierlingh tries to convey the idea that we should not neglect the art when dealing with tides. The expression *'use of time'* is an excellent one. This advice to *use* time is the one that is least observed. There are engineers who have enough patience but who cannot see several years ahead, or who will let a suitable opportunity pass. Vierlingh's 'patience' does not mean slowness or inactivity, it means choosing the right moment. He wants us to work

37

at the pace dictated by Nature and the tides. If a dam has been made and other works have to follow, wait and see the improvement yielded by the first dam. The 'use of time' is more or less 'the use of natural forces', of striking the iron while it is hot.

Vierlingh is an apostle of the offensive. The complaining tones of previous chronicle writers cannot be detected in his book. No ancient theories about God's incomprehensible anger resting permanently upon the country. He puts his finger on the weak spots. What, for instance, caused in 1532 the 'drowning' of the vast territory of Reimerswaal? Vierlingh had visited the dangerous breach in the dike near Reimerswaal in its early stages. It could have been repaired then, says Vierlingh, but the Lord of Lodijke, who had his estate near Reimerswaal and was not friendly disposed towards that city said: '*Oh, let the little harbour scour!*'

'But the little harbour is scouring still', says Vierlingh in his old age, 'and the Lord of Lodijke has lost all his beautiful estate, and the whole country is lost!' Reimerswaal itself followed soon after Vierlingh's death, the city vanished completely and the land has so far not been reclaimed. People gathered human teeth on the sandbanks and sold them to the dentists in the towns.

A little delay, a tiny quarrel, and there goes the property of thousands of hard-working farmers and burghers! There goes the labour and endless perseverance of so many generations who had faithfully maintained the Golden Hoop.

Vierlingh's comment is: 'The foe outside must be withstood with our *common* resources and our *common* might, for if you yield only slightly the sea will take all'. There was no other way, only by painstaking unity of all the people could the foe outside be kept outside.

Look at the much-afflicted island of Schouwen (Zeeland)! Superstitious folk say that in former years a mermaid was caught and that the merman, her husband, had said:

> '*Schouwen you shall rue*
> *That you have kept my frue*' (wife).

Vierlingh says that it is quite natural that almost half of that large island has been lost. If they had withstood the sea in the beginning, they would have been able to keep their island intact. They allowed the cancer to develop and now are unable to stop it. 'The more you retreat on rear defences, the more the sea takes the opportunity to throw you out of your land.'

'Tidal streams are like green boughs', he adds, 'you can curb them while they are still young. They are like naughty children who must be educated in their youth.' Even in our own days of science and machinery there are still many opportunities being lost by failing to see future dangers, or, seeing them, not acting while there is still time.

In quite recent years we ourselves tried to imitate the almost lost art of making new fertile land from sandflats, and found how sound Vierlingh's advice is. We tried with some *high* sandbanks where the plants were certain to grow. Yet, as we had not the experience of Vierlingh our disappointments were many. Vierlingh himself could not leave a sandbank alone, even those which scarcely rose above low water! Of course no plant can grow at such a low level, but by making small dams in the form of a simple cross and making these higher every year, he obtained small patches above the level where the plants would grow. There are always two quadrants in the cross where the lee of the dams will cause silting, and these quadrants, once they have silted high enough, will bear plants. And then behold! You get a new island which can be extended.

Vierlingh is a friend of ours, keeping us company when we walk over the wide tidal sands. 'The growing plants like to be looked at', he says. 'It is they that must give you the land that will ultimately yield great crops. It must not be a stiff clay, it must be a soft and sweet earth. The eye of the master makes the horse fat, and in the same way the eye of a great dike master causes new land to grow.'

'I have strangled many deep creeks', he muses another day; 'in the Lost Deep the currents were so fierce that our cables snapped as if they were made of glass. If such work has to be done, let nobody interfere.'

'At Hendriksland it came to pass that when we wanted to dam a creek, the opposite shore melted away as if it were butter on a hot iron, but this was a trick of the mattress makers. They wanted me to buy more willow twigs. Then I ordered them to begin from the opposite shore and like a lamb the creek submitted to strangulation.'

Vierlingh is a meek and religious man: 'It is not really such a great art', he says, 'a shepherd might be able to imitate it. But making new land belongs to God alone. For He gives to some people the intelligence and power to do it. It takes love and very much labour, and it is not everybody who can play that game.'

He hates official humbug and has some bad names for puffed-up dignitaries: 'The slippers, the nightgowns and fine fur-coats have no value outside the dikes. They must be men used to hard work from childhood, men who have greased leather boots on their legs and who are able to withstand a rough climate. Because in time of storms, rain, wind, hail and snow they must be able to persevere.'

From the attack upon the fine fur-coats and nightgowns we can gauge a point in Vierlingh's character which is well-defined. He loves his work and he knows that he is a good dike master; he is full of zeal and energy; fighting streams and tides is a great joy to him – but there is that human factor in the form of some small 'Big Man', who hardly knows ebb from flood, who, when put on the sand flats cannot say which is East and which is West, nor which is North nor which is South.' These dignitaries are accustomed to give unintelligent but high-handed orders from their elevated position of self-conceit, they lead a soft life and they will spoil every good plan. It would only be right if they were put on the tidal flat 'in jack-boots up to the point where they split' and made to walk the whole day long, like any ordinary workman, through the mud.

'I remember a Senator from The Hague who was so fond of pepper cakes that he never was without them and who had peahen's eggs in his trouser pockets to hatch them out there. This pepper-cake eater, etc . . .' He becomes really sarcastic when he starts talking about such highly-placed men, he loves far more to see his dear plants grow, or to 'strangle' some roaring stream. Preferably one whose 'roaring can be heard at a distance of some miles'.

'Your foe Oceanus does not rest nor sleep by day or night, but comes suddenly like a roaring lion, seeking to devour the whole land. To have kept your country is a great victory won. God has given us by His Grace the materials to fight the sea. They are willow-twigs, grass-turfs, stones, clay and straw and we must use them intelligently and subtly.'

'I saw the danger of the land' – he was called from his bed in the middle of the night. 'The gaps were sharp and dangerous, because the dike had fallen into a deep breach. In the early morning I made the drum go round the country, calling every creature to work. Those women who had no barrows had to carry clay in their aprons, others had to make hand-barrows, others carried clay in sacks upon their shoulders for lack of barrows. I got many people there and succeeded in repairing the gaps. Yes, he who has not seen the anger of old Neptune, how ugly he is in his grim wrath, should not complain when much land is actually being lost.'

Goethe would have liked to be present at such a scene, as the words he puts into the mouth of the dying Faust show, literally:

> Inside a land like paradise.
> The flood may rage outside up to the top.
> Green is the field, fertile. Men and cattle
> On this newest of soil feel comfortable.
> And when by storm a breach is made
> Common effort hastens its repair.
> Such a teeming company I crave to see.
> On free ground with a free people I long to be.
> I would say to that moment:
> Stay, thou art so grand!
> The tracks of my existence
> Then would not vanish into eternity.

In writing these last passages of his 'Faust', Goethe, himself in his last days, saw as his ideal a free nation striving for its existence on a free soil, keeping the evil waters out by united effort. He must have had the Dutch people in mind: the paradise behind the sea walls, the floods raging up to the very tops, and a teeming people on this newest of soil fighting against the elemental power of the sea and emerging victorious. The tracks of these poldermakers shall not vanish as long as people live in the polders they made. They may be the envy even of a Faust!

Everybody knows the tale of the Dutch boy who saved his land by putting his hand upon a mole-hole in the dike. This may be a fable, but Vierlingh's tale is true when he relates how during a very critical stage of some dike work the sea broke in again, rushing anew into the polder. 'But immediately a stalwart Frisian threw himself into that hole and the other people came quickly to help and so this stalwart Frisian saved the land.'

5. *Windmills*

There was water outside and there was water inside the 'Golden Hoop'. Both had to be fought vigorously. The farmers called the inland waters the 'waterwolf', because the many lakes in the country were gradually eroding their soft shores of peaty material. This was a process which the farmers disliked intensely.

Some of the lakes were natural, most of them had been made by the people themselves when they wanted peat to burn in their houses, either for cooking of for heating. Every town had its own lake where the peat was dug, and those lakes grew in size not only because of the dredging of the black substance, but also because of the attacks of the waves on its shores. The larger the lake became the greater the attack of the waves. The big Haarlem lake, for instance, grew to such an extent that Amsterdam itself was threatened.

The low-lying swamps of central Holland came late into cultivation. One can easily guess the dates of reclamation of the different regions of Holland by observing the pattern the ditches make. The first reclaimed areas are like jigsaw puzzles, having crooked ditches and irregular parcels; the later-developed parts have straight ditches, close to each other; the latest development shows a chessboard pattern.

The quaking soils never could be made fit for bearing crops, but the surface of them could be made to bear cows. The water level of those quaking soils is higher than in the lower polders. It seems paradoxical that the higher land should be the wettest and worst drained, but the reason is that when the water level is lowered by better drainage, the land itself lowers to the same extent, because the layer of wet peat on which the surface rests is very thick. It may shrink to a small fraction of its original size when it dries up.

At Kinderdijk, near Rotterdam, windmills, pumping the surplus water out of the country, are still abundant.Streamlining and ball bearings are the modern means to make them more efficient.

(Photo J.G. van Agtmaal)

People not only wanted fuel for their hearths, but also salt in their food. When the peat of those districts which had been flooded by the sea was burned, the ashes contained salt. This was called 'selberning'; a bad proceeding because the country was broken up into pools by doing so. When it was learned that the dikes succumbed sooner near such broken-up land and that the foundations of the dikes were thereby weakened, the selberning was forbidden. Nevertheless, much damage had already been done, especially in Zeeland. Had the peat digging continued, the country would have acquired the appearance of a large lake with islands in it. After the burning of our woods we burned our soil.

The great loss of land due to the inroads of the sea and to the extension of the lakes caused the people to wend their way to the sea, where they became traders and fishermen. It also put them on the track of inventing a means of pumping water out of the country – the *windmills*.

These windmills solved the serious problem in an elegant and economic way. They were our *first machines* and they marked a new era. The first wind-driven mill for pumping water out of the country seems to have worked in the year

42

FERTILE NEW POLDER

New land obtained by planting marine grasses or by pumping the water out of lakes gives high production. This picture shows a former lake bottom. *(Photo K.L.M.)*

1408[1], but it was not before the 'time of trouble' was over, i.e. about 1600, that it was used on a large scale. After that year the windmill provided thousands of acres of new fertile land inside the sea dikes by pumping the water out of the many lakes. It eliminated the threat of the so-called 'Waterwolf' and it also turned Holland into an industrial country. For not only could the wind-driven machines pump water, but they could also move the wheels and saws of a great many factories. When Napoleon came to Holland he saw to the North of Amsterdam a village, Zaandam, lying in a very watery district, with 860 windmills turning their sails. 'Sans pareil', said Napoleon.

The sluices in the sea dikes were an important invention in their time; they were clever, self-supporting devices, draining off at ebb, closed at flood. But the windmills were the instruments which turned a threat into a boon; they were enthusiastically hailed once their value became obvious. Since about 1600 more and more 'polders' in the Low Countries have been kept dry by continuous

[1] The windmill for grinding corn was probably an Eastern invention, but was employed here before the year 1000.

43

pumping. To-day interruption of this pumping would spell ruin to about half of the Netherlands. When the Germans left Holland in 1945 they took with them some key-part of each pumping station. But the inhabitants had foreseen this and had their duplicates safely stowed away, so that no damage was done.

A good exponent of this mill era, our first mechanical era, was *Jan Leegh-water*, who lived from 1575 to 1650. He was born in the midst of persecution – his family belonged to the Protestant sect of the Anabaptists, who would not bear arms and were persecuted more than any other. His birthplace lay north of Amsterdam in the midst of many lakes, separated by narrow streaks of land.

As Vierlingh is one of the experts in hydraulic engineering outside the dikes, so Leeghwater is one of the experts on hydraulic works inside the dikes. His name means Low Water. Perhaps, because of his enthusiasm to drain lakes everywhere, people grew accustomed to call him so. At least he used this name himself, often adding 'mill constructor and engineer' as his profession.

In those early days of mechanization a mill constructor was a man who could make use of the elements, a wizard who could 'take the wind in his fists'. The fairy tales had become, true, the goblins and pixies did the work overnight and the owner could stand and look at the work with his hands in his pockets. The farmers said that it was not horses, nor men, but 'John Wind' who perform-ed the miracle of grinding riches. Being a mill constructor was a modern profes-sion in those days, a profession for handy and ingenious young men.

Jan's father could remember the building of one of the first windmills north of Amsterdam. The farmers of that time had shaken their heads and said: 'What is this thing! We shall never again find an egg in the meadows.' But Jan himself adds: 'Instead of eggs we have now fat cows, also horses, also butter and cheese, and an abundance of good crops; all through the Grace of our Lord, by whose mercy an egg can hardly be weighed against a bean.'

About 1640 Leeghwater wrote two small but famous books. As was the case with Vierlingh, who belonged to a former generation, Leeghwater wrote down his experiences in his old age. Both had been great men in the national trend of work. Leeghwater's fame had then spread all over Europe. He had been called to France and England, he had given his advice in Eastland (Germany and the Baltic States) as well as in Denmark. His value was recognized by kings and nobles who wanted to make their country as fertile as Holland had become. The Prince of Orange needed him when he besieged the town of 's-Hertogen-bos, held by the Spaniards. The later had inundated the surroundings of that town, but Leeghwater put a great number of small mills, driven by horses, to the job and thereby the Prince was able to capture the important town.

Vierlingh's book shows that he was seized by the idea of teaching future generations the high art which he had mastered; he must have felt that there was a danger of this art being lost. Leeghwater is less grand, he is the inventive,

44

energetic, self-made man; eventually he sat down to write his experiences, which give vivid descriptions of the activities of a Dutch engineer in the 17th century. In his old age he was still an active propagandist for draining and reclaiming the country. He supported a great plan – the drainage of the Haarlemmermeer, the largest of all lakes in Holland (40,000 acres) – and his

LEEGHWATER

One of the famous draining engineers whose advice was sought after all over Europe; an expert in mill-making and in pumping water out of lakes. His name means Low Water. In his old age he wrote a book about the draining of the Haarlem Lake, the biggest of all Dutch lakes (40,000 acres). He wanted to use 160 windmills for it.

book was received with such enthusiasm that it had to be reprinted seventeen times within a short period.

Leeghwater shows in his other book how the face of the country changed in a few decades. In his home country, the peninsula north of Amsterdam, he had already counted by 1640 twenty-seven lakes which had been pumped dry, not including several small ones. In this book he also writes about his personal experiences. One of these reveals the difficulties of the ancient dike masters when working in a foreign country; he relates it as follows, in sober straightforward words:

'In the year 1634, a day before All Saints, I was in Eastland (Germany) at the new seawall of Butsloot (North Sea coast of Schleswig). I was the engineer and general surveyor there. In the evening a strong wind accompanied by

45

thunder came from the southwest. I went for a chat with Pieter Jans, one of my men, a Frisian, who had to make a new lock for ships. But the wind blew so hard that Pieter Jans said: 'Baas, remain here for the night'. But I said: 'No Pieter Jans, your house is no more than 5 or 6 feet above the field and my lodgings stand upon a dike which is 11 feet above the field. So I went to my lodgings.

'When I got there my son Adriaan and I went to bed in our clothes. The wind grew stronger and veered to the West, so that sleep did not come to our eyes. After an hour the waves were already hurling against our dike, very dreadful to hear. A little later, one of my surveyors, Sieuwert Meinerts by name, came banging at my door and cried. 'Leeghwater, come out of your bed now!' This we did. We took our coats and walking sticks and went towards the Manor which stood about 80 yards farther on, also on the top of the dike. Whereupon Sieuwert Meinerts said: 'You will be dead before you come to the Manor; the wooden beams are being blown from the pile so that you don 't know where to hide'. But we advanced at great peril creeping along the dike towards the Manor. The water had risen to the very top of the dike.

'And when we arrived there we found 20 refugees, men, women and children, all Dutch. Also 18 Eastlanders arrived, so that we were 38. The door was burst in by one of the waves and the water came into our boots. The Eastland women said: 'Dike Grave, och Herr Dike Grave, where shall we go?' But I said: 'We are all in the same boat.'

'My son stood in the kitchen and said several times plaintively: 'Father, have we to die here?' And this was very hard for me to hear. Whereupon I thought: 'Shall I die here with the Poops[1], when I have been in so many countries? That would be very hard indeed.' 'Whereupon I consoled the people and said: 'I hope that Almighty God will change all this.'

'And this lasted until three o'clock in the morning. We could not see whether the water had fallen or risen by then. The northern part of the Manor became undermined and collapsed, the money chest went with it through the floor. Also the southern part of the Manor gave way.
'When the day broke, the whole camp of the Dutch dikeworkers had gone with all the people in it.'

Leeghwater goes on to relate that a large fertile island nearby, bearing 23 or 24 churches, went as well. Seven or eight thousand people were drowned there. 'The house of Pieter Jans where I had been the previous evening had disappeared too with the whole family. My lodgings where my son and I had been had likewise been swept from the dike.

'Now when we had had this flood, we remained a day or two in the Manor on the dike. But then the Poops came and pursued us. They thought they would

---

[1] Poops = nickname for Germans.

get the dike master and all the other bosses of the work in a trap and so they came sailing with a ship towards the Manor. But we had also a very good ship from Medemblik in our own country, which could sail a little faster, and so we reached Husum in safety.'

A fortnight later the 'Poops' turned up again and then Leeghwater and his troop fled to the Duke of Holstein, who received them in a friendly way.

Leeghwater did not succeed in repairing these destroyed dikes. The work had to be abandoned in 1648, that is, shortly before his death. The job was too difficult, because of the fierce tides which ran in and out. But other Dutchmen took over the work and succeeded. The large clay-marsh islands of Pellworm and Nordstrand, whose population had been reduced to one quarter in that night of All Saints, when 44 breaches had been torn in the dikes and nearly all the houses and churches had been swept away, owe their existence to the energy and capability of the Dutch dikers who followed in Leeghwater's footsteps.

Leeghwater's Haarlemmermeer Scheme included 160 windmills. One of the sentences in his book is: 'The draining of lakes is one of the most necessary, most profitable and most holy works in Holland'. But the time for this great and very necessary work of draining the Haarlemmermeer had not yet come. The water could have been pumped out of that large lake even in those days and the man in the street showed great enthusiasm for it, but the mighty Waterschap (Polder Board), in whose domain the lake was located, saw too many difficulties. The first scheme to drain this lake had been put forward in 1617, but it was not drained before 1852, after 15 successive schemes had been proposed by prominent engineers in the course of 225 years. It was not drained by windmills but by steam power.

Leeghwater is one of the main pioneers, however. One of the three pumping stations of the Haarlemmermeer polder bears his name. The well-known aerodrome Schiphol lies at the bottom of that drained lake, which is about 5 yards below the ordinary high-water level of the sea.

6. *Ancient Dutch Abroad*

On the Eastern banks of the Elbe estuary, in the Wilster Marsh, we can find an almost complete example of Dutch scenery. Not only the marshy land but the drainage system, the farmhouses and the windmills are to a large extent identical with those of the Dutch province of Friesland. This is an example of one of the first Dutch settlements known in history. Frederic, Bishop of Bremen, as early as 1103 had invited settlers from the Low Countries to reclaim marshes in the neighbourhood of that city. Possibly there had been similar migrations even earlier. The Wilster Marsh was reclaimed by Dutch settlers in 1130. It is but one of many 'Hollandries' in Western Europe.

If reliance can be placed on old legends, the ability of the Dutch to turn bad land into good was already in evidence before Norman times. Ditches to drain the country were made before dike building began. The saga relates how a certain Walfridus and his son, who had been great men in draining parts of the low Wapelinge – they have always been pictured with spades in their hands – were slain by the Normans, while praying in the church of Bedum. This is a good example of a saga which has had power to survive. The power lies in the dramatic vision; a clear light is thrown upon the awful centuries when the holy work of draining and developing the country was hampered by the heathen raiders from the North. The vision is unforgettable; the sin of killing pious farmers who drain their land and who pray in a church of a holy village (Bedehem = house of prayer) is unforgivable.

Land drainage, land reclamation and dike building could not properly start before the Norman invasions slacked down, about 950. But soon a great advance was made, at home as well as abroad. Chroniclers relate in laudatory terms the drainage work undertaken by the Dutch Crusaders in the Nile delta. Dante, in one of his poems, pays a tribute to the Flemish dike and drainage people, another tribe of the Low Countries. Long before any written record, there was the Frisian settlement in marshes along the eastern coasts of the North Sea, the coast immediately south of Denmark, which has since then been called North Friesland. The North Frisians still speak the Frisian language in some form. They were described by Adam of Bremen in 1088 as able-bodied heathens, who jumped over the creeks and ditches of their marshes with the aid of long poles, so that nobody could get at them. Since then their country has been largely destroyed by the sea.

What kind of trekkers were they who went away for ever, singing the joyful song which has come down to our age: 'To Eastland will we sail, To Eastland will we go'? Saxo Grammaticus, a Dane of the 13th century, describes them in much the same way as Adam of Bremen: 'they are rough-minded people but agile of body, they despise the heavy and cumbersome cuirasses and fight with small shields, throwing javelins; they surround their fields with canals and swing themselves over these with the aid of poles.' – The Abbot Emo calls his countrymen impetuous by nature (1224).

The great trek of these 'agile but rough-minded' farmers towards Eastland[1] countries had started long before Saxo visited their country. It lasted until quite recent times. The people of the western shores, active groundworkers that they were, thoroughly versed in all things relating to the making and improving of soil, saw the gates of the hinterland open to them. The kings of those far-away countries invited them to come, offering them the low marshes, where no other

[1] Eastland = what is now called North Germany, Poland. Russia.

48

people would or could live, but where they throve. In a few generations they turned these wildernesses into so many paradises. Mostly they obtained 'eternal privileges', such as exemption from taxes or freedom from carrying arms. The paradise of the old-day farmers was without taxes and without arms. There were spades, ploughs and much work in God's open fields but no compulsory human commands.

Conquering a large part of Eastland would have been easy, but the Dutch, like the Venetians and the merchants of the Hanseatic League, were conquerors of trade, not land, and never gave a thought in that direction.

But the healthy exodus of emigrants, seeking freedom from paying taxes, from war and religous interference, continued throughout the centuries. The ancient saga of the Pied Piper reminds us how a stranger, dressed in queer-coloured, outlandish clothes, by his haunting tunes induced the children of the West to come to the East. There they vanished 'into the mountains', never to return. In those days Prussia, or Russia, or the Balkans, where according to the saga and geography the children went, were very far away and the parents who saw them go did not expect to see them again. They were 'lost'.

After the Middle Ages the trek continued, as the religious persecutions of the 16th century (the 'time of trouble') drove many of the proudest families abroad. Most of them were Mennonites, those stubborn Anabaptists who refused to carry arms. On the eastern shores, for instance in the Vistula delta, they were received with open arms, because of their skill and quiet zeal in creating fine land from swamps and bogs. German writers acknowledged their achievements with astonishment. The towns and villages are still there as evidence of the labours of these Mennonites and other Dutch people. The town of Prussian Holland in this delta was founded as early as 1297.

The trek of the Mennonites can be followed until 1860. Before and after the first world war many trekking Mennonites left Russia, where they had migrated from the Vistula delta by way of Poland. They had their own university there, and after 1918 made their way to the United States (Kansas), and to the Gran Chaco in South America, where they started making new settlements once more. The Holland-America Line brought them without payment to South America, because they had lost all in the revolution.

Besides this trek of anonymous masses to Eastland and other countries such as South Africa, where the Dutch farmers founded the Boer States, or to the Eastern shores of North America, where they created the colony of New Nether-lands with its capital New Amsterdam (now New York), the history of the 16th and 17th centuries contains some outstanding names. They include our first hydraulic engineers, such as Leeghwater, Vermuyden, De Wit, Van der Pellen, Meyer, Van den Houten, Rollwagen, Van Es, etc. These hydraulic engineers worked in practically the whole of Europe, from Sweden to Italy and from

49

England to the Volga. When we read the books about their achievements we cannot help admiring their energy and courage. Even with modern means some of the works they attempted would be outstanding. Yet, most of these intrepid men died in misery. None of them grew rich. The good they achieved was for future generations, not for themselves. They were the heralds of the new and first mechanical era. They could make and use machines as well as their brains. Their most outstanding qualities next to their engineering abilities were their spirit of enterprise and their dogged perseverance.

It was a kind of venturous, romantic energy which characterized the old Dutch engineers. They had the technique and knew that their inherited art could help other countries. The bread and life of countless generations was assured if the marshes of Russia and other countries could be made fertile in the same way as had been done in Holland, in the Vistula delta or in one of the other 'Hollandries'. Yet these engineers, capable as they were, were alone in a vast foreign land and their works must needs be of long duration. In the beginning they had the power of some far-seeing King or Tsar behind them, but once such a progressive ruler died, or war broke out, their plans were abandoned and the engineer was forgotten. Another constant obstacle was the opposition of the peasants. Drainage of marshes meant change of work for fishermen and fowlers. More than once they cut the new dikes in the night, thus destroying the work of years.

In Italy, for instance, the Pontine Marshes near Rome attracted Dutch enterprise. It started with the 'Dic-Maestro' *Gilles van den Houten* in 1623, while later *Nicolaas Cornelis de Wit* took over the work. They had acquired the monopoly from Pope Urbanus VII, but a friend said aptly of De Wit: 'he who wishes to catch two hares at the same time gets nothing'. De Wit wanted to do too much at the same time and did not succeed. After him came *Nicolaas van der Pellen*, but he too had to give up the work, in 1659. Then *Cornelis Janszoon Meyer* came and studied the Pontine Marshes carefully. His book: 'Del modo di seccare le Paludi Pontine' is still well known. On his death in 1701 the work was not quite finished, but his son carried on with it. In 1707, however, the inhabitants destroyed all the work, cutting the dikes which had been built so laboriously. In 1720 the whole region was marsh again. An entire century's work had come to nought and the Pontine Marshes had to wait until recent times, when they were reclaimed by Mussolini. The creators of the ideas died, the ideas themselves lived on and were carried out when the time was ripe.

In England more success could be achieved than in Italy. One of the Dutch engineers who worked in England was *Vermuyden*, whose works are described briefly by Edward Cressy in his 'Outline of Industrial History', as follows:

'Vermuyden had the good fortune to enlist the aid of successive Stuart sovereigns to his schemes. Before his time the Fen-district was a vast marsh, 2000 square miles in area, and open to the sea. The old churchmen, who held the monastries on the islands of the Fens, made numerous efforts to reclaim the land in their districts; but work of this character had been so seldom carried out in England that no man had the experience which would have ensured success. The Government took action, officals were appointed, . . . but in spite of all these precautions the Fens were for the most part unreclaimed down to the end of the 17th century; and the fenmen eked out an unhealthy and precarious existence by grazing, fishing and fowling.

'The great authorities on the control of water in those days were, of course, the Dutch, whose country had been wrested from the shallows of the sea. It was only natural, therefore, that when, in 1621, a serious breach occurred in the Thames wall at Dagenham, James I should invite one of the Dutch engineers to effect the necessary repairs. Having accomplished this work to the satisfaction of his royal employer, Cornelius Vermuyden drained the Royal Park at Windsor, and reclaimed 70,000 acres of land at Hatfield Chase in the district of Axholm, where the island of that name rose above a waste of inland waters. Meantime, the kingly patron was no longer James I, but Charles I.

'A local jury was summoned by the King to consider the question, but they broke up, after expressing their opinion of the utter impracticability of carrying out any effective plan for the withdrawal of the waters. Vermuyden, however, declared that he would undertake and bind himself to do that which the jury had pronounced to be impossible. The Dutch had certainly been successful beyond all other nations in projects of the same kind. No people had fought against water so boldly, so perseveringly, and so successfully. They had made their own land out of the mud of the rest of Europe, and, being rich and prosperous, were ready to enter upon similar enterprises in other countries.

'It was a condition of the contract that Vermuyden and his partners in the adventure were to have grafted to them one entire third of the land so recovered from the waters.

'The task was carried out mainly by the aid of Dutch and Flemish workmen and met with great opposition from the fenmen. Over and over again the embankments were broken down and drains filled up, while the conflict led to loss of life and the invocation of law. But Vermuyden was undeterred by the difficulties which beset his path. He rose superior to every untoward circumstance, and having carried his project to a successful issue, he turned his attention to the West Country.

'In 1631 the problem of the Fens of the Eastern Counties was attacked by Francis, the Earl of Bedford. The services of Vermuyden were obtained, and an energetic start was made. Again the fenmen banded themselves together, and hindered the work in every possible way. Much was made of the fact that it was being undertaken by foreigners, in whom no desire to benefit Englishmen was to be expected. The situation was rendered worse by the fact that the district was not so easily drained as Holland, so that mistakes were made[1], and by the shortage of funds from which the promotors suffered. The King then intervened, and the scheme was revised and extended. But a political crisis was at hand. Civil war broke out and interrupted the work for many years. The Earl of Bedford died, and when the final and successful attempt was made, it was his son who was concerned, and the Commonwealth which celebrated the achievement.

'But the price of tenure of the newly-won land was unremitting vigilance. The retaining walls had to be maintained in a sound condition, the drains kept clear, and the water pumped from the lower levels by the force of the wind. So long as these matters received attention the aqueous tyrant was subdued and turned to useful purpose. Six hundred and eighty thousand acres of the richest land in England which had lain beneath its surface were brought into cultivation, and a dismal, unhealthy swamp converted into a healthy and fruitful plain.'

Quite recently Mr. L. E. Harris of Cambridge, discovered what has been thought to be the portrait of Vermuyden, together with many other portraits of Dutch 'adventurers' who had provided part of the money Vermuyden needed. These portraits are in the possession of Colonel Noel of Worcester, one of the descendants of the 'adventurers'. Mr. Harris wrote a new (1950) study about Vermuyden in which this remarkable character was more or less analyzed. There is not much of a personal nature in the documents, but one decided conclusion can be drawn, namely: 'That he was a man of boundless energy and tenacity of purpose'.

In 1949 a Bill was passed in the House of Lords to improve the river Great Ouse – which drains the Fen Country – at a cost of £ 6,250,000. There had been several schemes before, but this last one was decidedly the best and cheapest. The remark was made during the discussions in a special committtee of the House of Lords that this scheme, which seemed so modern, appeared already in broad outline on an original drawing of Vermuyden of 1642. Vermuyden apparently had not been able to finish his whole scheme, only the western part of it was executed. This proves how sound Vermuyden's ideas were, notwithstanding the lack of instruments and formulae. His son was one of the Colonels of Cromwell. Many of the money-providing 'adventurers' seem to have become great men in England; they married into the nobility.

The generation which had to do with Vermuyden made rhymes with some trace of annoyance:

---

[1] I doubt whether such mistakes were made. v. V.

The Dutchman hath a thirsty soul,
The drainers are up and a coile they keep,
And threaten to drain the Kingdom dry.
Why should we stay here and perish with thirst?
To the new world in the moon away let us go
For if the Dutch colony get thither first
't Is a thousand to one but they'll drain that too.

But this has long since turned to hymns of praise.

The 'simple, full-bred fenmen', also called the 'Yellow Bellies', because of the malaria, destroyed Vermuyden's work, stood by with loaded guns, killed his men, 'and swore that they would stay until the whole land were drowned again, and the foreigners forced to swim away like ducks'.

One of the Dutch capitalists wrote as early as 1628: 'The mutinous people have not only desisted from their threats, but now give their work to complete the dike, which they have *fifty times* destroyed and thrown into the river'. Later came the real opposition; what the captalist was speaking about was only a mere prelude.

Many dikes and drainages were carried out in France. But in Germany, Poland and Russia the 'Hollandries' were the most abundant. Along the Molotschna there were 46 Dutch villages in 1836; the district Chortitza had at that time 20 such villages. In 1900 there were about 10,000 Russian Dutch settlers in America (Kansas). In Poland there were about 2,000 villages inhabited by the descendants of the Dutch immigrants, in Posen there were 830 villages.

The first great canal in the United States, the Erie Canal, was financed in 1792 by the Dutch and its locks were devised by Dutch engineers. Until 1798 the United States of America had no other creditor than Holland.

7. *Balance of Losses and Gains*

The success of the dikemasters and farmers who worked to gain new land outside the sea walls has been remarkable. Since the earliest reliable records, i.e. about 1200, the area of the Netherlands has increased considerably, and Nature, as if wishing to recompense the incessant toil of men, has produced the most fertile soil imaginable.

Roughly speaking, two-thirds of the lower part of the Netherlands is manmade, while the other third part is 'natural' sea marsh or moorish swamp. Since about 1200 we have gained:

On the sea shores	940,000	acres
By pumping lakes dry	345,000	,,
By pumping the Zuiderzee dry . . .	550,000	,, (partly future)

In all   1,835,000 acres.

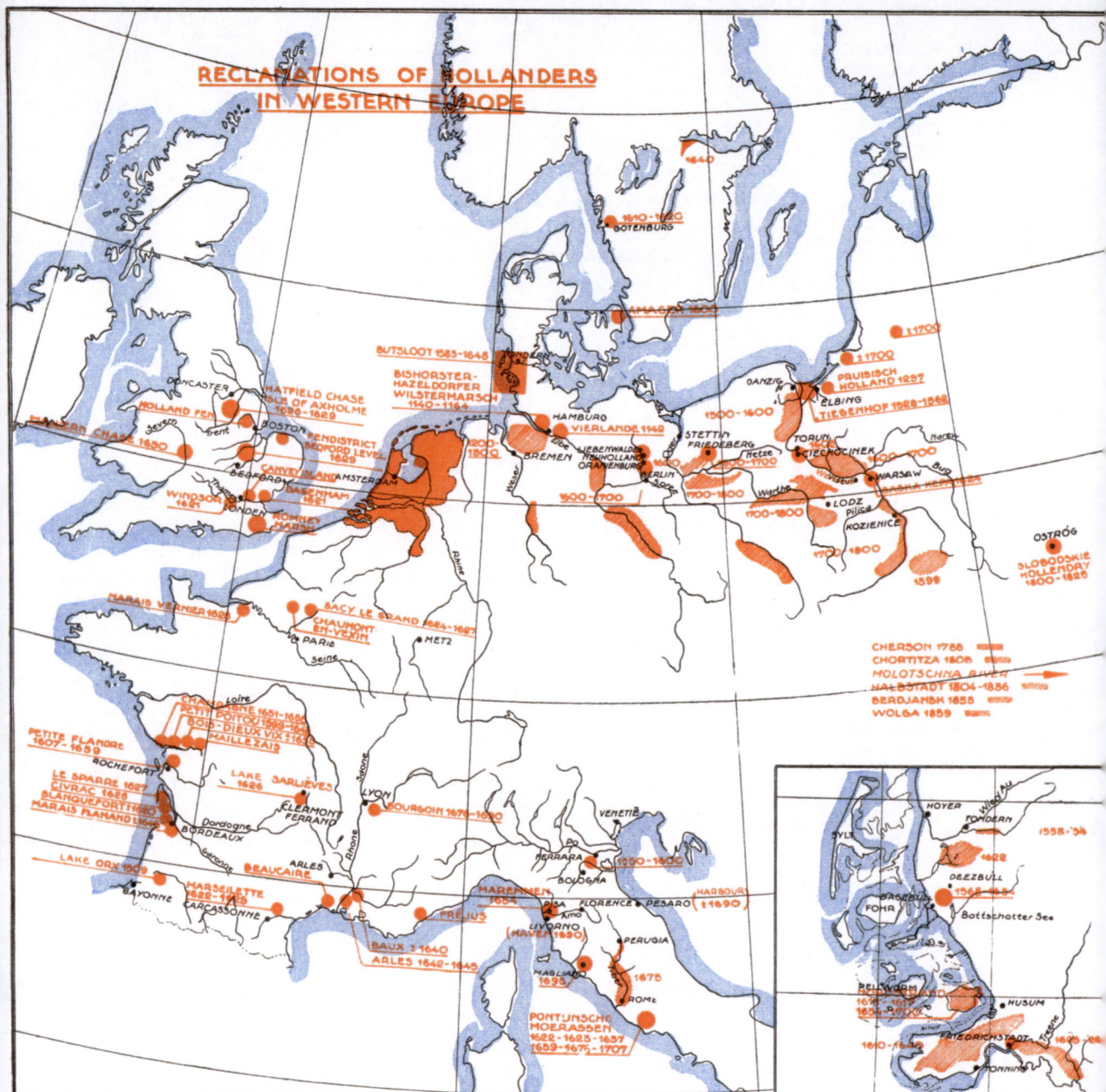

Western Europe (1955) – the field of activity for Dutch farmers and engineers since 1100 and earlier.

Right-hand page:
Works in Russia were mainly carried out after 1700. The many Dutch colonies in Russia are not mentioned here. *Van Suchtelen* was Director of the Russian Topographic Institute and 'the almighty porte-parole of the Czar, a man soft and learned, but timid and without nerves, – a first-rate financier'.

Wollant was Director-General of Transport in Russia. He built the Ladoga Canal (190 miles + 110 miles river canalization), Leningrad-Volga Canal (6 locks, 50 barrages), Beresina Canal (100 miles, 14 locks), Oginsky Canal (33 miles, 8 locks), King's Canal (16 miles), the Dnieper rapids canalization (41 miles, 9 locks), etc. He was 'an admirable and untiring man, a real godsend', according to the Russian Prime Minister.

Cruys was Vice-Admiral of the Russian Fleet, made charts, fortresses, mills. The name means 'cross'; he used a white flag with blue cross for the Lutheran Church he built in his garden. This flag is still the Russian Naval Ensign.

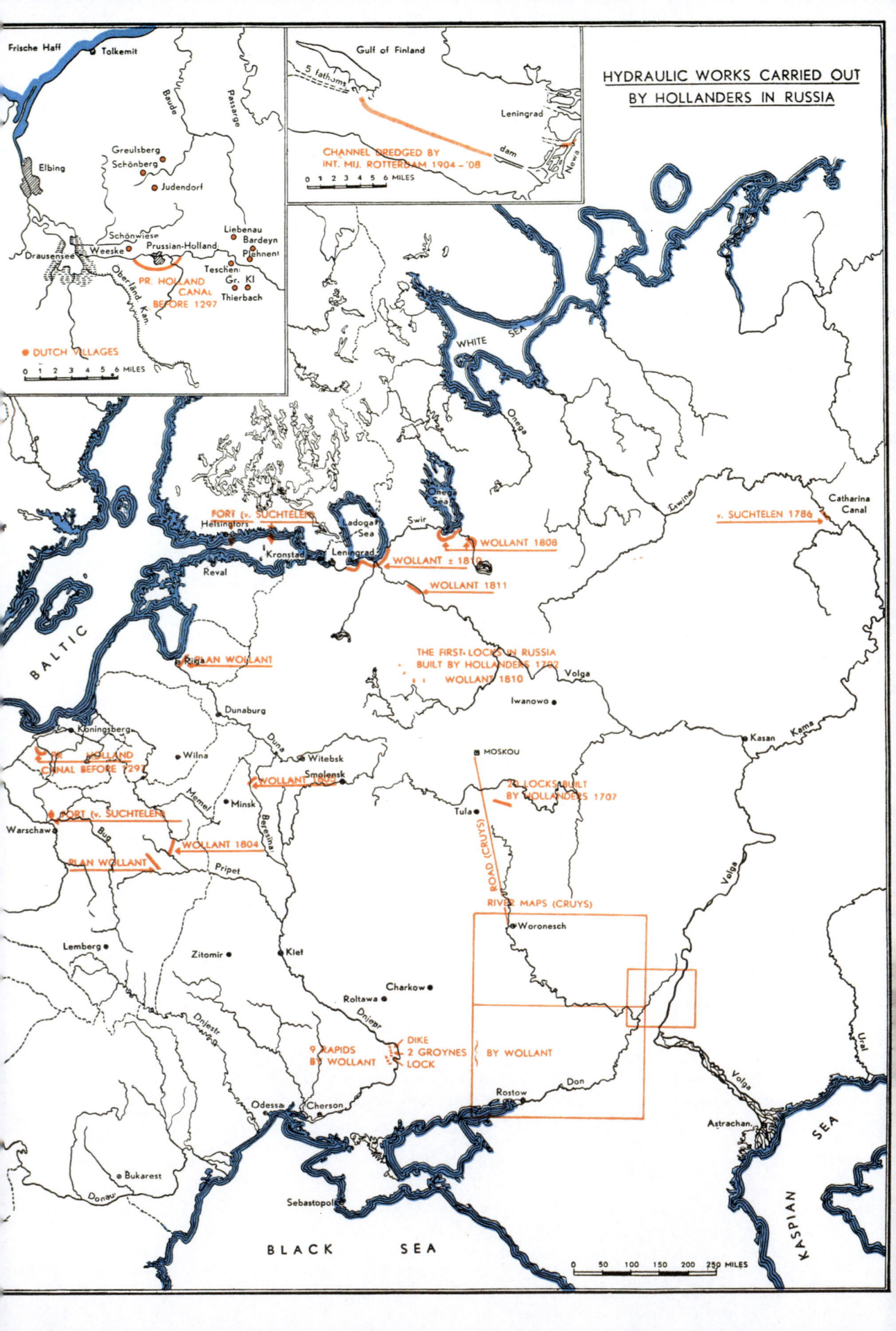

We must compare those 940.000 acres recovered from the sea with the losses we have suffered since 1200. These losses, about 1,400,000 acres, exceeded our gains. This account will be adjusted now we have set about the task of draining the last polders of the Zuiderzee.

The southwestern part of the country – Zeeland – has had an adventurous history. Much land was lost there and after centuries much was recovered. As one of our poets aptly puts it:

'Show me the people – who can understand? –
'Living where land became sea and sea became land!'

The well-known coat of arms of Zeeland, a lion half emerged from the sea and bearing the motto: LUCTOR ET EMERGO (Wrestling I emerge) is well chosen. When the lion has totally emerged from the water, the coastline of this region will have decreased from about 500 to 60 miles.

A general picture of the ups and downs of the history of the Netherlands may be obtained by studying the graph showing the man-made areas gained every quarter of a century outside the sea walls. In the beginning our gains were not remarkable. Much new land was made and surrounded by dikes, but those dikes were not strong enough to last. After about 1500, more land was gained *and held* when men like Andries Vierlingh energetically encouraged the growth of plants which retained the silt and the dikes were made stronger. In the first quarter of the 17th century even as much as 80,000 acres were gained *and held*. This record has never been surpassed. It was achieved partly by the reclamation of much of the land lost during the preceding wars. It should be noted that the great activity in land-making during the last half of the 16th century coincided with the most relentless of persecutions, known as the 'time of trouble', during or before the 'time of Alva'. In those years thousands of men and women were burned to death because of their religious beliefs, but these troublous times are not revealed in the graph, nor the Eighty-Years War (1568–1648), fought with mighty Spain, the enormous Empire on which the sun never set. The good work on the dikes continued, war or no war. The dikes of about 1580 were even far higher and bigger than they have been ever since.

About 1600 began the new era called 'the Golden Age'. It was the century in which Holland fought for her freedom and position with one hand and with the other gained mastery in the art of making sea defences with such success that no great losses of land have occurred since, while much new land has been made by the silting method. With a third hand we undertook the task of pumping the lakes dry, which had threatened the country from the inside. And not content with these labours, we developed an unprecedented shipping trade and became one of the greatest colonizing nations. It was as if some dam had suddenly burst. The pent-up strength and prowess of the nation had risen

56

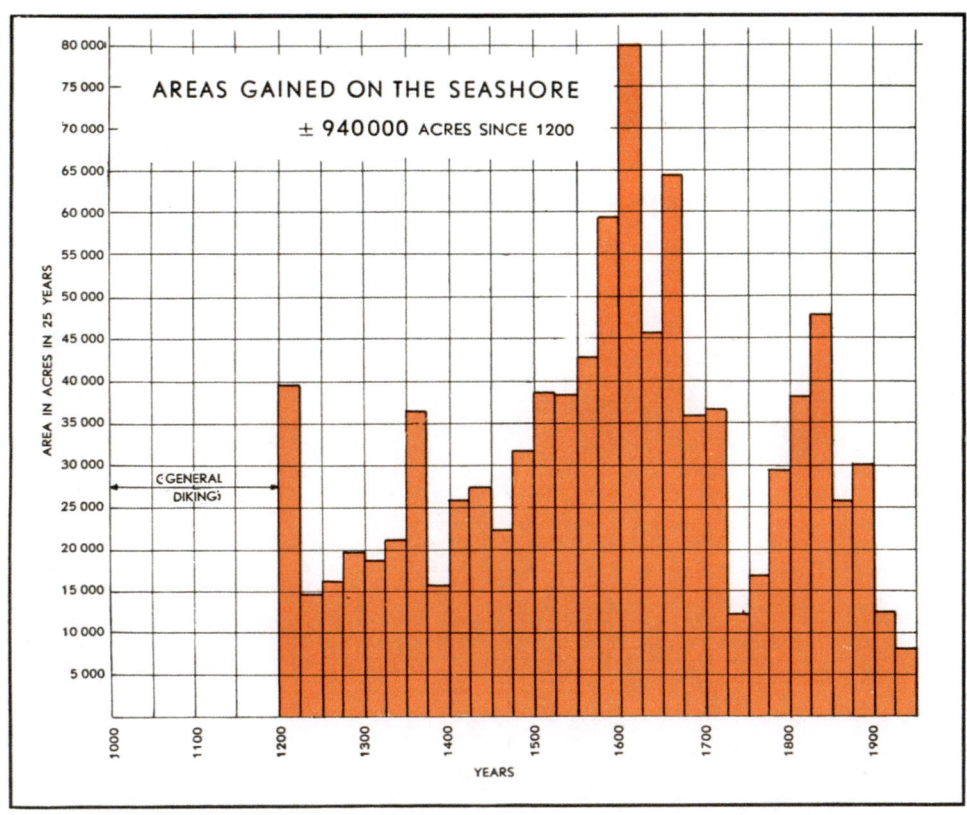

AREAS GAINED ON THE SEASHORE

± 940000 ACRES SINCE 1200

AREA IN ACRES IN 25 YEARS

(GENERAL DIKING)

YEARS

By the patient process of planting marine plants and making small dams much fertile land has been
gained on the sea shores.

higher and higher and now poured forth triumphantly over the sea. It was the
reward of 20 centuries of severe education in freedom, reclaiming, colonizing,
shipping and trading.

About 1600 *Sir Walter Raleigh* reported to his Queen that the British had 300
vessels in the Baltic trade, whereas the Hollanders had 3000. Altogether the
Dutch owned as many ships as eleven European nations combined, Great
Britain included, he said. *Colbert*, the famous French statesman, wrote about
1650: 'The sea trade of Europe is carried by 25,000 ships. Of these, 14,000 or
15,000 belong to Holland and only 600 to France.'

To William the Silent everything had seemed lost in 1582, when he spoke of
the necessity to persevere, even without hope of success. Then, a few years
later, came this sudden incredible success, proving how steadily the strength
of the farmers, skippers and traders had been increasing for centuries.

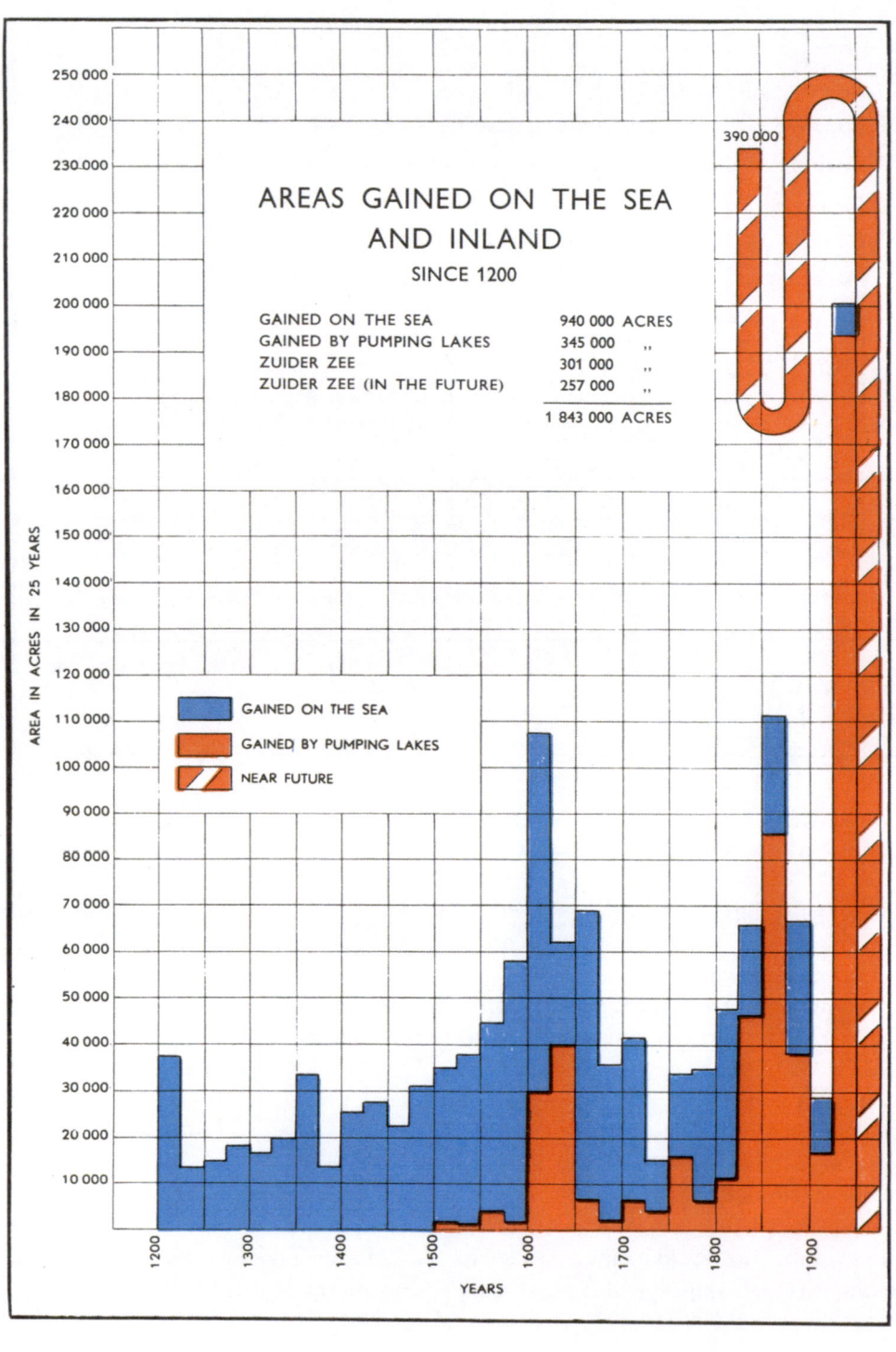

AREAS GAINED ON THE SEA AND INLAND

SINCE 1200

GAINED ON THE SEA	940 000	ACRES
GAINED BY PUMPING LAKES	345 000	,,
ZUIDER ZEE	301 000	,,
ZUIDER ZEE (IN THE FUTURE)	257 000	,,
	1 843 000	**ACRES**

390 000

AREA IN ACRES IN 25 YEARS

- GAINED ON THE SEA
- GAINED BY PUMPING LAKES
- NEAR FUTURE

YEARS

An ebb followed this outburst, perceptible – inter alia – in the decline of the area of land gained from the sea. What were the exact reasons for this weaker growth? Had the 'slippers, the night-gowns, the gorgeous fur-coats' and the man hatching peahen's eggs in his trouser pockets got the upper hand, as Vierlingh had feared? – They had.

After the Napoleonic wars, and even before them, there was a good recovery in land-making, but this could not be maintained either. The making of new land outside the walls of Holland, the work in which Vierlingh excelled, has decreased considerably in recent times.

The reason for this decline has nothing to do with degeneration, however, but with the change in economic circumstances. The industrialization of the 19th century – 40% of the Dutch are engaged in industry now – and the vast plains of North America took much of our available labour. Money could be made more quickly by other means than by the slow process of land accretion. But now prospects are better. The great increase of our population necessitates the gaining of more and more land, and the State has realized its function. Before 1850 it had been private enterprise which made the new land, afterwards the State became – had to become – interested, slightly at first, later more strongly. It is understandable that a private person cannot go beyond the limit of direct personal profit, but the State can and even *must* go beyond this principle. It must try to increase the *national income*.

That is the difference between '*private economy*' and '*public economy*'. A network of canals or roads can no longer be financed on private economic principles with tolls, nor can land reclamation be paid for in that way.

The Dutch Government first showed that it understood this new idea about 1918, when the Zuiderzee project was accepted. 'Private Economy' never could have found the financial balance for the great work, but 'Public Economy' leads to the execution of great plans by which national income will increase. It does not neglect 'private economy', but it goes further.

In 1880 this important lesson had not yet been learned when a rich man, *Teding van Berkhout*, wanted to increase his wealth by making a dam to the Frisian island Ameland in order to gain much new land. The Government did not encourage him, or oppose him. It only said: 'The dam is likely to break, because you do not make it high and strong enough'. But Teding van Berkhout answered that he did not have the money for making a more robust dam and pleaded for help. When the dam was finished and shortly afterwards actually broke during a storm, the Government would not assist, and as the man's money ran out, the dam remained broken and was totally lost. The result was, labour

---

The difference in colour indicates the difference between old and new methods. The new method is gaining land by pumping; the Zuiderzee bottom could only be drained by first making artificial lakes.

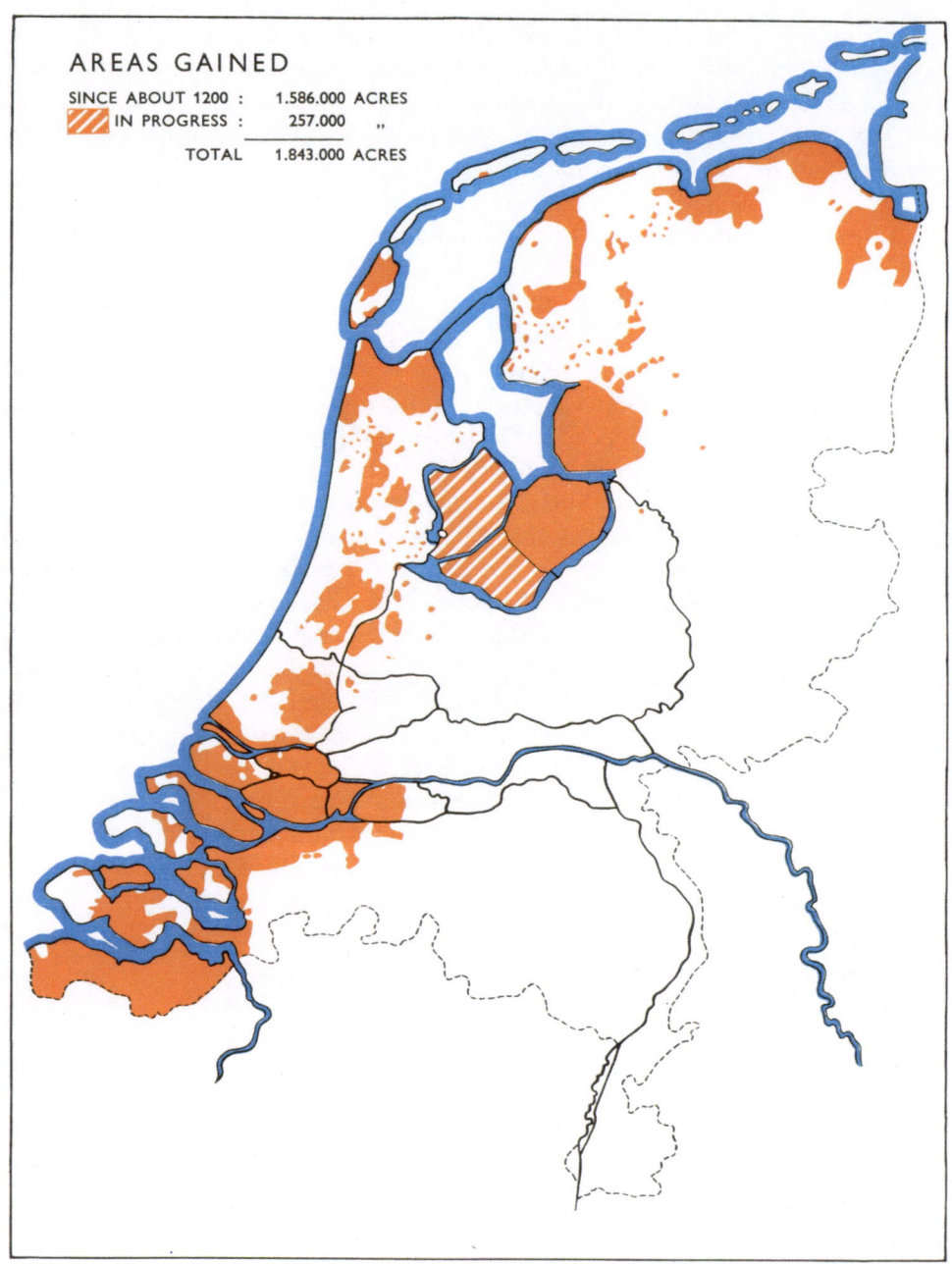

AREAS GAINED

SINCE ABOUT 1200 :	1.586.000	ACRES
IN PROGRESS :	257.000	,,
TOTAL	1.843.000	ACRES

The red spots indicate the areas which have been obtained either by accretion, or by pumping lakes dry.

lost, money lost, no road to Ameland, no land gains. The Government should have helped because of public interests.

There had been another lesson before. In 1852 the Government had actually drained the large Haarlemmermeer, a work too large for private enterprise. But the Government had done this work very timidly because the financial gains by selling the lots of the new polder were slightly less than the initial expenses. No villages, schools, churches, farmhouses, roads, locks were made, nor other necessities, the polder therefore became a wilderness in which the two first generations led a miserable life; bankruptcy, sickness, death, immorality and failure followed. Because of lack of good drinking water even cholera occurred, and the death rate was exceedingly high for some decades. It was the third generation which reaped fruits.

Large plans cannot be viewed from too narrow principles. The Zuiderzee works have proved that the making of good, fertile, *inhabitable* land costs more than its selling worth. So does a harbour, a road or a canal. From a private-economic standpoint the making of the Zuiderzee polders therefore has been a loss (to the present generation); from a public-economic standpoint it is a gain (for the future generations). For such schemes three reports are wanted:

1. Technical report (plus cost estimation).
2. Financial report (private conomic, or how to find the money for carrying the scheme out).
3. Economic report (public economic. This last report must show whether the scheme should be carried out or not).

The Zuiderzee works prove that our present generation is not so sluggish as the graph of land-gains outside the dikes might suggest. If we study the graph showing the total accretion of land, inside as well as outside the dikes, we get an entirely different impression. This graph shows that our generation is the greatest of all in making new land. That is as it should be, for to-day we have machines infinitely more powerful than the tools of old. Yet, the method which Vierlingh used so successfully 'at low cost' has been neglected for at least a whole century. The State will take up Vierlingh's methods again.

Exact figures as to the amount of soil which our ancestors removed in creating the country cannot be given, but rough estimates can illustrate the magnitude of the task which has been carried out.

We have already estimated that in the early centuries some 100,000,000 cubic yards of earth were carried to the artificial hills which are called 'terpen' or 'wierden'. Those hillocks were made only in a small area, covering roughly 8% of the country. We compared this work with the largest of all the pyramids, which had a capacity of 3,500,000 cubic yards. It might be added that the Suez Canal, as constructed by De Lesseps in 1867, had the same content of 100,000,000 cubic yards.

The sea walls or dikes were our second work. In 1860, that is just before the advent of steam dredging, we had about 1750 miles of them, containing about 200,000,000 cubic yards of material. Moreover, there were many old deserted dikes, whose contents may be estimated at 50,000,000 cubic yards. Those 250,000,000 cubic yards were practically all transported by handbarrows, wheel barrows and horse-drawn carts.

The third great work was the digging of the ditches and canals. In the lower half of the country about 800,000,000 cubic yards of earth have been removed, in order to drain the land and separate the fields. Of shipping canals there are about 4800 miles in Holland, for which a figure of 200,000,000 cubic yards would be a fair estimate.

The fourth and greatest task was the digging of peat. This digging served a double purpose: the provision of fuel and the creation of lakes which, when drained, gave more fertile land than the original moors themselves.

In total we have dug according to this rough estimate the enormous volume of some 10,000,000,000 cubic yards. This includes the making of lakes as well as the digging of moors in the higher eastern regions of the Netherlands.

Compare this figure with the dredging of the Suez Canal. We constructed about 100 Suez Canals of the size made by De Lesseps. All this *was done by hand*, whereas De Lesseps used 60 steam dredges.

To date, the dredgings of the Suez Canal can be estimated at about 300,000,000 cubic yards. We may say that before the year 1860 we had dug (by hand) about 33 modern Suez Canals. This is equivalent to a ship canal of 40 feet depth, 200 feet width at the bottom and a toatal length of 5000 miles, a distance from London to Calcutta. The Queen Elizabeth could easily sail along such a canal.

The foregoing figures do not include all our soil work. Much sand has been removed from the dunes and brought into the moor districts to fertilize them. The roads were built with sand, often carried from afar. The farmers wanted a level surface for their fields and therefore removed large quantities from one part of their fields to the other. There was also earth work for the foundation of houses and other buildings, for improving rivers and for making new land, and the constant re-digging of canals and ditches. I think our Suez Canal would become very long indeed if all those quantities were to be added.

The energy that went into this spade work! Sitting behind a typewriter, we shudder to think of digging a paltry cubic yard or two a day, because clay is a heavy and coarse material. An old dikemaster told how three of his men had dug 132 cubic yards of tough clay in one single day. This was a top performance. The diggers who remove the artificial hillocks, called wierden, in order to use the earth as a fertilizer, reach the large figure of 76 cubic yards per man per day! Yet the engineering handbooks consider 10 cubic yards a fair daily average for one digger.

8. *Three Williams the First*

The feudal system of Emperors, Counts, Dukes and other Lords was not indigenous in northwestern Europe. It was rooted in Roman administration as the names imply: Kaiser, Czar = Caesar; Prince = first; Count = comites, friend; Duke = dux, leader. The Frisians of the old days in the North of the Netherlands, whose diking history was followed in the opening chapters, could never get accustomed to lords. It would be false to say, however, that the feudal system only meant suppression; it has had its definite advantages. We could not remain in the simple state of rural communities. Foreign influences of this kind had to be digested properly and feudal lords have often proved to be a blessing: not all feudal lords forced men to shoot apples from the heads of their sons. To a large extent the Swiss and the Dutch wrenched themselves free from extreme feudalism at an early date.

The low coastal plain aptly called the Nether-Lands, which stretches northeastward from Blanc Nez (the White Cape opposite the cliffs of Dover), gave birth to three successive centres, after its original peatlands had been almost wholly destroyed by the sea. All three were anti-feudal.

The *first centre*, the Frisian, sprang up to the North of the lagoon, which was named Zuiderzee by the people of that centre, because the name means Southsea. It was pre-Norman and rural. There were no towns, but many dwelling mounds show that it was well populated. The new siltations after the destruction of the peatlands made living possible.

The *second centre*, post-Norman, i.e. after 1000, was Flanders. The name seems to mean 'new tidal accretions', or 'saltings'. It had, and still has, important towns like Ghent, Bruges, Antwerp, but it was no less democratic and anti-feudal than the Frisian centre.

The *third*, again very democratic and anti-feudal by nature, grew up slowly and late into what now is called Holland-Proper, or Randstad-Holland. All around a Central Moor, which the Dutch call Holland (Holland-Proper) villages sprang up, many of which developed into towns almost touching each other. If a line is drawn from Dordrecht over Rotterdam, Schiedam, Delft, The Hague, Leyden, Haarlem, Amsterdam, Hilversum and Utrecht, that line encircling the Central Moor country, shows the place where about half of the population of the Kingdom of the Netherlands live. That large moor with its Ring-town Holland, inhabited by about five million people, could not have developed without large engineering works.

By a whim of Nature this Central Moor between Amsterdam and Rotterdam,

The Hague and Utrecht, has not yet been destroyed by the encroaching sea. When directly after the Romans left the sea level rose slowly, a Claw of the Sea moved in from the south near Cape Blanc Nez and took the peatland area, where later Flanders and Zeeland were to emerge. Only a few poor, naked, low and salty islands were left.

By 1330 this Claw reached Rotterdam, by 1421 Dordrecht and Gorkum.

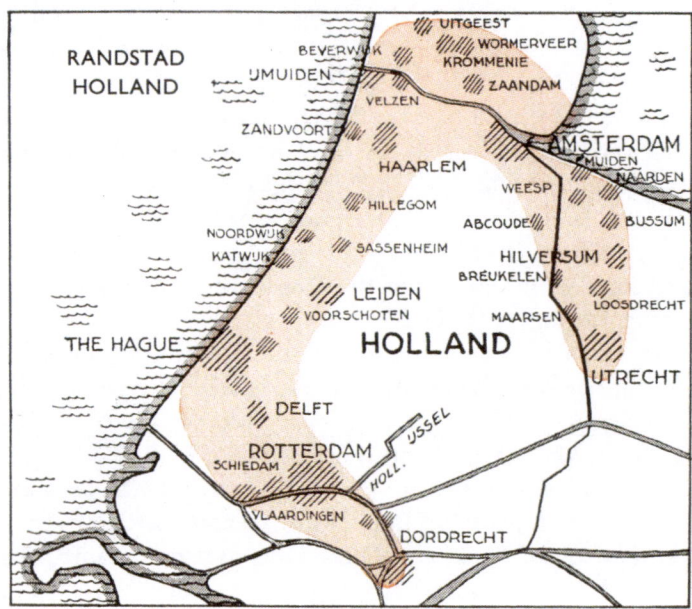

The so-called 'Randstad Holland', comprising the towns of Amsterdam, Haarlem, Leyden, The Hague, Delft, Schiedam, Rotterdam and Dordrecht, population about 5,000,000. It grew around the big central moor, called Holland-Proper: first a moor; then a conglomeration of lakes, because the moor was burned in the stoves of the Hollanders; and after 1880 a fertile field, because the lakes were pumped dry.

Earlier, another Claw of the Sea penetrated the peat plain from the North, and about 1250 it had come as far as Amsterdam (not then in existence) and Haarlem. The distance left between the two claws was now only 37 miles, which is the diameter of our still-existing Central Moor.

Small as it is the Central Moor has never been taken by the sea, . . . not yet. It even became the modern centre of the Netherlands. How did this evolution come to pass and how was it possible that this extremely vulnerable country has never known a catastrophe, while all the other parts of western and northern Netherlands had many?

About 800 a Moorish traveller passed through this centre-to-be. He came from the sea and wanted to visit Utrecht, a city of Roman origin. Disdainfully

64

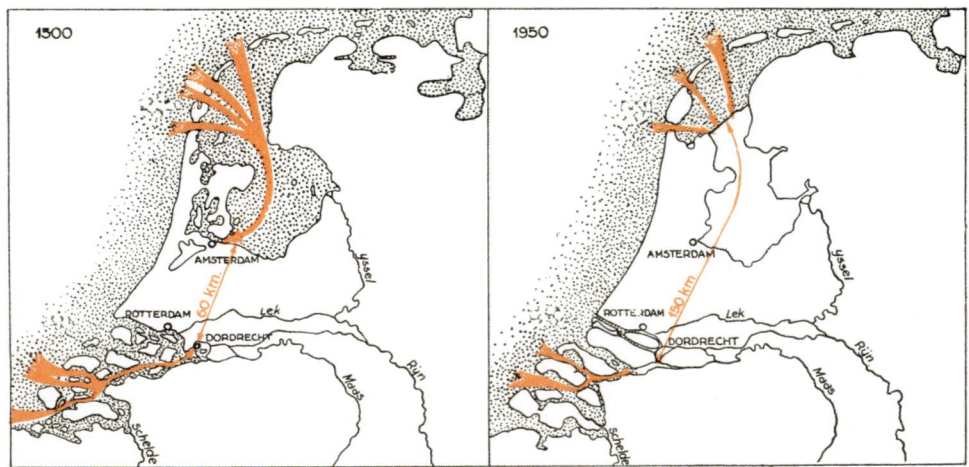

Several claws of the sea, of which two were extremely dangerous, threatened to destroy the lower half of the Netherlands. The claws advanced bit by bit until about 1400, when they were checked by the strong dikes which Count William I had built around Holland-Proper. From 1421 to 1932 there was an equilibrium between the forces of the sea and human resistance against them. In 1932 the main northern 'claw' was pushed back by building the Zuiderzee dam. The southern 'claw' began to be pushed back since 1950 but in 1953 proved to be a still-formidable threat.

he wrote: 'The land is a sebscha there' (a saltwater and mud plain). Of course, it must have been waste and almost uninhabited, stormfloods could enter and kill fresh-water animal and plant life. There were only a few inhabitable places; to the seashore a ridge of sand dunes in whose shelter some Frisian farmers lived. Along one of the branches of the Rhine, called the Old Rhine, some inhabitants dwelt whose farms were of the Saxon type. Near Leyden, on the clay deposits of the Old Rhine, a house existed, called *Holtland*, and here some historians presume the family Van Holland lived, the family that gave the Central Moor and its sheltering dunes a long series of Counts. Their names were mostly either Dirk or Floris. The name Holtland means woodland.

As has been stated, feudal lords such as Counts and Dukes, were not easily accepted .The Marsh Dutch were farmers and skippers, the 'boers'[1] accustomed to rule themselves. They disliked Counts with their tolls and taxes, wars and ambitions. They did not want to be 'possessed' by some Count, and be called 'dear subject'. They said they were born noblemen and born freemen themselves. But there were too few of them in the Central Moor and its sheltering dunes to oppose the Dirks and Florisses.

As early as the year 1018 Dirk van Holland defeated an army of the Emperor of the Holy Roman Empire, his own Sovereign who had bestowed the title of

---

[1] 'boer' means builder. He 'builds' crops, he 'builds' the soil.

65

Count upon one of his forefathers. The imperial army had been sent to chastise Dirk; henceforth the Counts of Holland were practically sovereigns. They had to make their living, however, and though their county provided fish and fowl, it was none the less but a poor domain for human habitation, consisting of little more than water, peat and a dune ridge. They therefore established some tolls on their rivers, compelling tradesmen in boats to pay heavy dues. By such levies feudal lords lived and fought their wars. It was because of the imposition of these tolls, that a certain Alpertus of those days not unnaturally called the Counts of Holland robbers.

Slowly but steadily these Counts increased their influence but the land itself, Holland-Proper, remained a sodden, poor part of Western Europe, the actual existence of which was precarious. The floating peat, about 50 feet thick, might vanish any moment during a stormflood. It was definitely backward in technical development. To the north, on the other side of the Zuiderzee, was the old country of Friesland; well populated, its people were living in a modest safety on many artificial mounds, the Golden Hoop already built around their lands. To the south was Flanders, emerged by natural silting after the destruction of its original peatlands. By 1150 it had exceedingly thriving towns; already sheltered by strong dikes, it was the mighty industrial and trade centre of Western Europe. Holland-Proper, the part at the combined mouths of the rivers Rhine and Maas, was almost nonexistent. Of all the low lands between Cape Blanc Nez and Denmark, the area with a diameter of not yet 40 miles left over by the two crablike Sea Claws, Holland-Proper, was the last to be developed.

It must have been hardly worth while to make dikes around such a poor country, and it was evident that very much energy would have to go into such a work. There were not many 'fen-men' either to make the dikes. Moreover, the Central Moor was not in one hand; Counts and Bishops on the eastern side claimed parts of it.

By the laws of Nature the area of Holland-Proper should have been a second Zuiderzee. Or rather, the Zuiderzee should have had a southward extension. The tide should have entered in Zeeland and it should have run behind the dunes up to the gaps between the Frisian islands in the North. This has not happened. The Cinderella country, in medieval times on the verge of disappearing, still exists, it has even become the richest part of the Netherlands, its very centre, and one of the richest spots of the earth.

What type of man was it who saved that soaken and poor country at the right time? We never shall discover the names of those who made the Golden Hoop around the Seven Frisian Sealands – the Nines, the Twelves, the Unknown Asegas – but since a new study[1] has been made about this important man,

---

[1] Mr. S. J. Fockema Andreae: 'William I, Count of Holland' 1954.

we realize that *William the Diker* (as I shall call him) made the Golden Hoop around the Sebscha, late it is true, but just in time to avoid its destruction. And extremely successfully, as 750 years of history have shown. This very low and very rich area has never suffered a catastrophe due to stormfloods, . . . at least not yet. This is almost more than remarkable.

William (1168–1222) was a second son. His mother was the daughter of the King of Scotland. Those who may have wondered about the absolute similarity of the Scottish and the Holland-Proper coat of arms, may find a clue here. The boy was named after William the Lion, the King who employed two Dutchmen, Freskin and Fresekyn (both Frisians apparently), as Counts of Sutherland (1197) and Inverary (1204). Young William's eldest brother Dirk was to inherit the County of Holland, so William, the second son, was sent to Scotland to be the Count of Ross, that wildish district lying at the end of the world, which was his mother's inheritance. William the Lion liked to have kinsfolk from the other side of the North Sea to assist him in subduing his Highlanders. Young William was the first to come, a boy of about 20 then; Freskin and Fresekyn came later.

Willekyn, as he is named in the Scotch annals, was too young for that task, and when the boy heard that his father had gone on a crusade, 'the best prepared crusade ever seen', he left his important assignment and went to the Holy Land.

Who could blame him! He was of a family of rovers. Europe had no frontiers as yet, except those against the Moors. Grandmother Sophy from his father's side, had been four times in Palestine and had her grave in Jerusalem. His grandfather, his father, his uncles in the Low Countries and those in Scotland, were all crusaders; they were accustomed to roam over Europe on horseback, or to be in ships on the Atlantic and on the Mediterranean Sea. Long voyages of many years did not mean much to the Vikings nor to the Knights of the 12th century. The 'subjects' of his father had the same style of living, several of them either trekked to Eastland or they too wanted to fight the Moors. So William, now a young man of 22, joined his father to fight in the crusade. Floris, the father, was not pleased, as he had wanted William to make something of his mother's wedding gift, the County of Ross. The clash between father and son was such that the Emperor of the Holy Roman Empire, who himself was the commander of the crusade, had to intervene.

Old Emperor Frederic Barbarossa, then 70, probably laughed in his red-grey beard. He succeeded in pacifying the angry Floris. Soon afterwards Barbarossa slept with his fathers. Legend says that he never died, he sleeps in some German mountain. Floris too lost his life. Young William fought bravely on, hoping perhaps to create a position in the East for himself. He had forfeited Ross, Holland was for his eldest brother, so why should he return? But the crusade was not a success and after a stay of five years in the East he came back to his ancestral moors and the ridge of dunes. This was in 1195.

67

William, now a healthy, handy Knight Errant of 27, a brave warrior, a good leader, a man with 'sea legs', had all the qualities to become successful, but he was poor. He was given an income of 300 pounds a year, the revenue of one of the river tolls. His education had been excellent; he had been strengthened by the hard times in Scotland and in the Arab country, and he had seen the civil-

William I, the Diker, Count of Holland-Proper, 1168–1222. This Crusader and Engineer-General points at one of the weakest spots of the Dutch coast (Katwijk).

engineering works in Flanders, Italy, Egypt, Asia Minor and England. Also he must have been influenced by his mother, Ada, who spent her childhood at Huntingdon on the verge of the English Fens where priests had already reclaimed some parts successfully. Such a young man should not remain a Knight Errant.

One of the Seven Frisian Sealands apparently saw his fine qualities and allowed William to cross the Zuiderzee. He must have been glad to accept any job. The Frisians, those free 'boers', who had already killed one of his ancestors when the latter tried to invade, and were soon to kill two more Counts of Holland (and their mighty armies), distrusted Counts, especially Counts of Holland. No nobility even of their own tribe could spring up in Friesland. The Frisians said they were noble since time began, and they could prove this with ancient documents saying that Charlemagne and even the Romans had treated them as such. In very ancient days the Emperor had ordained that the Frisians should welcome a Send-Count at Sudermuda on the Zuiderzee. The sluice there has still the defiant rhyme in stone: 'The tide will take its round, it waits on Prince nor Count'.

But William was an exception and went there. In the six years that he stayed in Friesland, he did not behave like a bossy Potentate; he served that country as well as he could. He even made the dike, designed to protect a newly-acquired part (conquered from the Saxons) against the floods of the Zuiderzee, a work to be proud of. After this, in 1203, his eldest brother Dirk died, and William, now Count of Holland himself, returned once more to his home moors. In the nineteen years that followed he made a name that will always be revered amongst the Dutch nation.

We often wondered who was the master engineer who created the marvellous Great Holland Polder, south of Dordrecht, the work which had included the damming off of the tidal mouth of the river Maas, and the leading of that river into the Rhine. This proved to be William I. He had already finished that gigantic undertaking by 1213. The polder was destroyed in 1421 by the St. Elisabeth's flood, described in a former chapter. William was a man of great conceptions. He surrounded the entire area of Holland-Proper with strong dikes and made several canals intended to drain the vast moors. They also served as a splendid network of shipping canals. It is likely that he made the dikes around the Zeeland islands Walcheren and Schouwen too, and that he established the still-existing administrations for the upkeep of these islands. The other parts of his clever and amazing reclamation and construction programme cannot be described here, but it is very clear that he knew the geopraphy of his county by heart. No maps as yet existed!

He was a 'multiple purpose planner' and even more: he carried those plans out himself. His son finished the work, as William died before his important undertakings were wholly completed.

In William's time there were no other technical means but the spade and the cart. He used 'ban', the accepted and acknowledged power of a Count to make the whole masculine population work for the common good; a kind of conscription, it is true, but here a question of goodwill and reasoning. Also, it was of much importance to the success of his own endeavours that his father and brother had already done the preliminary work. He must have had in mind that Holland-Proper was a backward country, as compared to Flanders, Friesland and Italy. The ripe and human Knight who had travelled over practically the whole known world must have been filled with a desire to make his little country up-to-date; he ordered and no doubt inspired the farmers and carpenters to defend the land against the sea, and to develop it by drainage. Perhaps one of the reasons why he could do so much was the lack of clerks and paper. William had no time to make clerks write and troubadours sing about his glory. And, of course, he had not enough money to have a decent court; all the energy and money must have gone into his programme of civil-engineering works.

He had no 'court civilization' as was the custom with feudal lords. Sometimes he used the priest of Vlaardingen as a clerk to write down some verdict, but this occurred seldom. This lack of written material is the reason why we knew hardly anything about him and his work until 1954.

Neither was he a builder of cathedrals; he did not go into competition with southern Counts and Bishops who built the stupendous Gothic cathedrals of those times. Safety had to come first, and after safety the reclamation of the moors, a long and expensive programme.

From 1216 to 1220 he was a crusader again, commanding 212 ships. He conquered such places as Lisbon and Damiate, successful as ever. Abbot Emo describes this crusade in a very lively manner. William's life was sound and moral, full of work and adventure. Twice he had narrow escapes in Zeeland, when he had to hide himself among the cargo in the hull of ships. We see him as an admiral defeating the French fleet in Flanders, and later in London to help compel King John Lackland to sign the Magna Charta, thus helping to stem the tide of royal feudalism in England, introduced by William the Conqueror in 1066. He carried out his national programme in a short time, despite crusades, floods and wars.

William is described as physically handsome, and from what we know of him he must have had a charming personality. The family of his fiancée saw him unexpectedly when he was dressing; he was 'pulcher, audax, strenuus' – well-built, audacious, strong. After his first wife had died he married the widow of the Emperor, which shows what a poor but good and gallant Knight could achieve in those days. He must have passed a large part of his life in ships. When inspecting his dikes and canals there was no other way, and there is no better way to keep in touch with the peasants and learning about their needs than being on board the small ships which voyage on inland waters. He must have understood his skippers, fishermen and farmers very well, and they him. He even had not a fixed home; he, the engineer-general, had to be at the dikes with his men, living as they lived in small inland ships. This sober man of action seems to have travelled by sea, rivers, canals and roads all his life, directing his dike-men and sluice-builders, when not abroad for other important work. He resembles a modern civil engineer, but without the paper and committees.

He was the maker of the heart of the Netherlands and therefore the main cause that the present kingdom exists. About 250 years later this heart was already strong enough to free itself from Spain. It was soon to become the centre of world trade in the 16th century.

A purposeful man with an ambitious scheme and remarkable success. How could William I ever be forgotten for 730 years! His deeds should have sufficed to make posterity praise him for all centuries to come. He created a work of the size of the Zuiderzee works, making for instance the dams in the Amstel and in the Rotte where henceforth Amsterdam and Rotterdam could spring up. His grandson William II became Emperor Elect of the Holy Roman Empire.

Dutch history knows three Williams the First. There was the William referred to above, Count William I van Holland († 1222), secondly, Stadtholder William I van Nassau-Orange, the Silent († 1584) and lastly, King William I van Orange († 1842). All three strove for the technical development of their country and for its safety against floods. All three had much contact with other nations of importance, nations who at the time had something to offer. The three Williams were wise broad-minded men, perhaps the best Rulers we have ever had. Each of them stood at the gate of a new era, each of them opened that gate. *Stadtholder William I*, the Silent Prince of the Dutch revolt, had been the Governor of Holland -Proper, Zeeland and Utrecht for about 20 years before he took the side of his people against the Hapsburg Emperor Philip of Spain. As such he knew the conditions of the dikes by heart. He had witnessed the tremendous floods of 1530, 1532, 1552 and 1570. William's dikemaster Vierlingh, who writes about these floods, has been mentioned before. William of Orange, diplomat and younger friend of Emperor Charles at Brussels had grown up at the court to become wise, silent, mild and unsophisticated. The tendency of this man of 'Dutch descent' was far from autocratic; he would have despised the device of later feudal despots: 'l'état, c'est moi'.

Here was a clash of religion as well as a clash between ancient southern feudalism and ancient northern democracy. A rich Prince, William gave his riches to the cause of his countrymen he served so well in its revolt; his mother Juliana stood firmly behind him and so did his brothers. Some of them were killed in battle; Spain was an exceedingly mighty foe.

William, of course, did not forget the other foe which in the years of his lifetime had caused such unheard-of damage to his country and to his own personal possessions, the sea. In 1581, the year of the abjuring of Philip, William wrote a remarkable sentence showing how well he knew the Dutch character and how far-seeing he was. He admonished his people 'to look forward to the future, and take good care, that this splendid fertile land, which we like and care for so much, and which the human foes have not been able to subdue, either by force or by slyness, be not wholly destroyed by bad maintenance of the dikes, to its utter ruin, to the annihilation of its people and to the deserved reproach of our progeny'.

It must be acknowledged, the Hapsburgers Charles and Philip, though dictator-like exponents of the feudal system (especially Philip), had taken very good care of the Dutch dikes. Never have we had higher and more massive dikes than those of 1580. In Friesland, the Spanish Governor Caspar de Robles, strengthened the dikes greatly. His name is still honoured in the Netherlands, as is his retort to the parsimonious provincial authorities: 'Please, try to close the breaches in the dikes with your ancient letters of freedom from taxes. Pave your dikes with them and wait for the next storm!' When Philip was denounced by the Dutch in 1581, the danger existed that the Hapsburg centralized administration for dikes would topple down.

William of Orange apparently saw this danger and tried to prevent it. He failed, however, as he was assassinated in 1584. No hope of success had appeared on the horizon for him. After his death the dikes of the Netherlands were neglected. Formerly the touching of any dike with a spade had been punished with death. Even the houses had been removed from the dikes that the latter might serve their all-important purpose and remain unmolested. Only sheep were allowed on them. Now, houses were built on their sides and in some places even cellars were allowed to be made in them. One of the main responsible bodies gave the example, it made deep wine cellars in the dikes, to be able to celebrate the conferences in the administration house built on the main dike of Holland. However, due to the strength of the dikes of 1580, aided by the kindness of Nature, a calamity has not occurred, . . . not yet.

We must feel sorry that the lifespan of William the Silent was not longer. He, who had ordered Van Deventer to make his splendid maps of the Netherlands as early as about 1540, could have prevented this neglect and shameful decentralization in dike affairs, because the people in their distress had given him all power. He, with his sharp intelligence, kind disposition, an open eye for the basic needs of his countrymen and an enormous personal influence, could have steadied the first faltering steps of the new state.

Likely he himself was the main promotor of the making of those heavy, safe dikes, which the Centre and the South of the Netherlands had about 1580, and which were so much higher than they are now. We know that he foresaw the possibility of the reproach of later generations and he was right in this. Those who will study the history of the dikes of Holland can find much with which to reproach those who came after William the Silent.

William influenced the Dutch character greatly by his service and behaviour. They are described in our national anthem, written in 'the time of trouble' at its worst. William speaks in it as follows: 'Here I am, all what I am and have done has been an endeavour to be 'a good instrument' of Right for the sake of my people'. It is not a prayer to save his life, nor a wish for glory, nor even a flaming attack on unrighteousness, but a description of what a man, 'free and fearless' should be and could be.

Every Dutchman, in time of personal or national need, remembers the lines about steadfastness, faithfulness, compassion, and the desire 'to expel that tyranny which wounds my heart so sore'. Our second William I should have lived some years longer. He saw, as did Vierlingh, that the fight against the sea was a *national* one, as was the fight against Spain. Long ago the human foe changed to become a friend of Holland. The other foe which William so rightly feared, the sea, still lies menacing and near.

Count William's statue, at Katwijk, points his finger to one of the weakest places in our sea-defences, with a lordly gesture. *Prince William* points to the governmental flaws regarding maintenance of dikes. Both say: *'Beware, this is no joke!'*

The civil engineering works of *William I, the King,* will be described later. He was raised in England in the early years of the 19th century where, after the Napoleonic wars, industrial inventions began to revolutionize the world. He fought bravely in foreign countries against Napoleon. No doubt it is due to these experiences that William had an open eye for our technical needs, and it explains why he was able to modernize the stricken country after 1815, when about half of the population of the towns of Holland had to live from gifts of the other half. He is called the King Merchant, but he might as well be given the title King Engineer.

All three Williams the First deserve the gratitude of the Dutch. All three were admirable men, servants of their country, peforming great tasks for the safety and the development of Holland. They provide inspiration for future Rulers and for future civil engineers throughout the world. The history of the Netherlands shows what part civil engineers may play in the welfare of a nation. Eventually civil engineering will make most Cinderella-countries fertile, rich and agreeable.

It is mainly a question of initiative and research. A good start causes more good starts to follow.

CHAPTER II

DREDGES' WORK

9. *'Waterstaat' and 'Waterschappen'*

Holland may be compared to a ship floating in the sea. The salt water is ever ready to ooze in.

A ship needs constant alertness and vigilance on the part of its crew. Inspections and repairs must be carried out lest leaks occur. No parts must be allowed to become rusty, the machinery must be kept in perfect running order. Even the most inaccessible nooks and crannies of the ship must not be overlooked.

As times passes on, the 'ship' has to be modernized too, or it will lose its income. Many are the mouths which have to be fed on our 'ship', more than eleven millions now on such a small artificial part of the earth. Holland is the most densely populated country in Europe and its birth rate is the highest; its death rate is the lowest; the population is increasing rapidly.

The authority responsible for the general upkeep of this is *Waterstaat*. The name indicates the state or condition of the waters, as well as the Authority of the waters. The task of Waterstaat engineers is to maintain up-to-date conditions in the country. Initiative and vigilance must be their main virtues.

What are *Waterschappen?* They are in charge of the watertight compartments of our ship, called Polders. There are more than 2800 of them[1]. These Boards are the direct descendants of the Asegas and Wise Men, who were chosen in bodies of nine or twelve to 'find' and maintain the Law of Right. In their time of small communities the Jurates, called 'The Nines' or 'The Twelves', had the power to decide life and death, peace and war, as well as having jurisdiction over the dikes, ditches and roads. But in the course of time these ancient rural councils, self-governed as they originally were, lost much of their independence and jurisdiction, except in the case of water and dike control. So the Asegas became Dike Masters. The control of the local waters and the defence against the sea has largely remained in local hands to this day.

A Waterschap is not a private society. It has autonomic functions, and the farmers who are chosen by their neighbours exercise power. It is true that the Provinces control the activities of the Waterschappen, but nevertheless the latter are self-governed in many ways, keeping their traditions and their responsibility.

A combination of the Governmental Waterstaat and the Provinces and the local Waterschappen may prove to be possible. It would mean centralization

---

[1] It should be pointed out that during the last few years there has been a move towards combining polder authorities in larger and stronger units. *S*.

Modern pumping unit used in draining part of the low country.

with decentralization. The Government might leave a certain amount of freedom to the people to have their own way, but keep a watchful eye on them to protect the general welfare. As regards safety against floods today the Dutch Government has practically no power. This seems a dangerous situation for a thickly-populated country below sea level.

It is one of the characteristics of the Dutch State that it has always had many free farmers' families capable of guiding a rural community in a democratic way. Men raised to respect blind obedience cannot be trusted as leaders of a small or large governmental board. Such a man might strand the ship. The ambition to serve one's native polder as a dike master or as a 'Heemraad' [1] has worked as a stimulant in building character and in developing a sense of responsiblity. Are these sufficient for saving the whole low country? It has been proved that they are not. Co-operation, study and capital are also necessary.

The long training in individual development and in collaboration for the common welfare proved to be a safe antidote against the subtle poison used by Nazis in our times, and formerly by other groups using inquisition in some form or other. This poison aims at killing a man's free spirit. It always comes

[1]) Heemraad = 'Home Warden'.

as a wolf in sheep's clothing. Like the threats of the sea, it is ever present, seemingly docile and sleeping, but suddenly it assumes an agressive form and the whole country is flooded with it in a disastrous way.

Most polders have had an exciting history. The archives are there to witness that it has often been far from easy to bring the polders safely through the dangers devised by nature or the human mind. But this has sharpened the wits of the dike masters and 'Heemraden' and has kept them alert. Is it not an honour and a source of pride to be found worthy to sit in the ancient seats of the Polder House and glance at the long row of names of the dike masters of bygone ages? Revered local names are there, of men who served their native soil faithfully in their time: their words and deeds can still be read in the Polder Books. The rich inheritance they bequeathed, the Polder, is the fruit of their labour and each generation must pass this inheritance on to the next in a better condition. Noblesse oblige!

Compared with the ancient Waterschappen or Polder Boards, the Governmental Waterstaat is a recent innovation. It took a long time to establish a Central Government in our country, and when it eventually came, it did not want to meddle too much with the intricate local water problems. In the 18th century appeared the first signs of the Central Waterstaat Authority which was to come; the time of scientific methods was approaching. Those with a scientific outlook were the first to propagate the idea of a Central Water Board.

In 1682 the Burgomaster of Amsterdam, *Johannes Hudde,* fixed the height of the sea level at Amsterdam by means of eight large marble stones. It was done with such a degree of accuracy that we could hardly have improved it nowadays. The sea level was measured 48 times a day accurately, later 24 times, and the books in which those figures have been noted down now occupy a shelf 100 feet in length. This was the beginning of scientific methods and the 'AP' or 'Amsterdam Level' became so famous that it was used over the whole of North Germany, where it is called 'NN', and even as far as in Poland. One of the marble stones of Hudde still exists; it is used to determine the geological subsidence of the country.

In 1726 a well-known expert in hydraulic engineering asked the Government for permission to study the water conditions of the country. He wanted to compare the water levels of the various polders with the height of the mean sea level and with those of the many sea and inland dikes; he also wanted to know the causes of scouring and silting of the rivers (how modern!) and moreover, how the defences of the coasts and dikes against the attacks of the sea could be constructed in the most economic way. In fact, he wanted to investigate all the problems relating to the physical conditions of the country. He would do all this for the moderate amount of £ 2500, but the money was not granted.

This pioneer of modern hydraulical research, who bore the good Dutch name

76

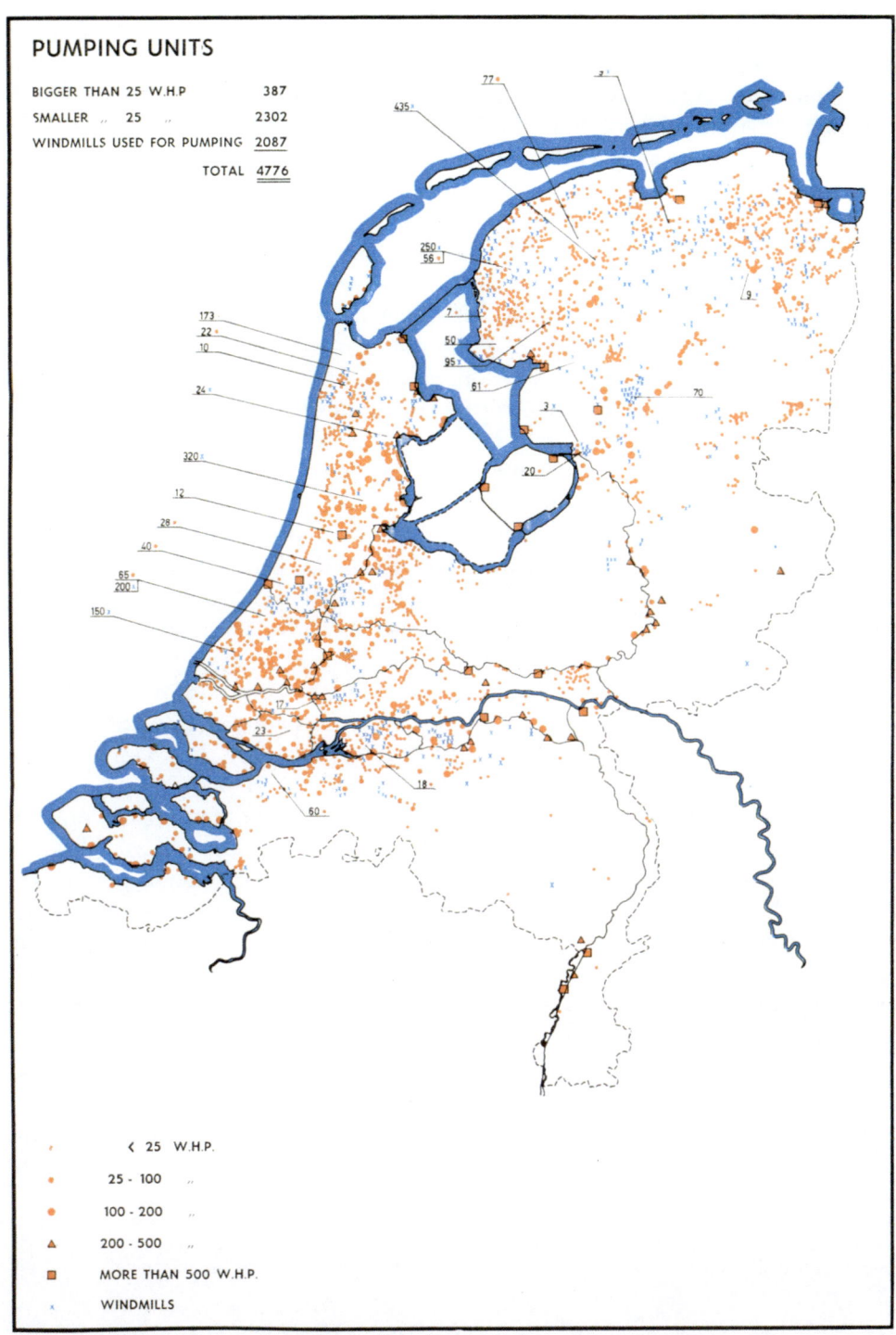

PUMPING UNITS

BIGGER THAN 25 W.H.P	387
SMALLER 25	2302
WINDMILLS USED FOR PUMPING	2087
TOTAL	4776

.	< 25 W.H.P.
●	25 - 100 ,,
●	100 - 200 ,,
▲	200 - 500 ,,
■	MORE THAN 500 W.H.P.
×	WINDMILLS

Holland is kept dry by continuous pumping with more than 4500 units.

of *Kruik*, but for some odd reason preferred to call himself Cruquius, was born too early. The idea of autonomy was still too strong, and the Central Government too weak. We had to wait till recent times for a Hydraulic Research Bureau to be inaugurated. Suppose Kruik had started this research as early as 1726, what a change this would have brought about in the safety of our country and in the whole world of hydraulic engineering!

Though Kruik's action did not have immediate success, the seed had been sown. Research, the Herald of Science and Improvement, could not be held back much longer and it found ways of circumventing excessive governmental inadequacy. There came into being private scientific bodies, which still exist to-day and which have remained active ever since their founding. Especially the 'Bataafsch Genoostchap voor Proefondervindelijke Wijsbegeerte' roused the public interest in the physical problems of the country and created the atmosphere of research by putting questions to be answered by anyone who wished to do so. If the answers were acceptable, prizes were awarded and those answers published. This is still done in our days.[1]

State organizations had become too old-fashioned towards the end of the 18th century. In 1754 the men at the helm of the Ship of State, probably influenced by the establishment of the 'Ponts et Chaussées' in France in 1750, had shown some goodwill by appointing an 'Inspector General of the Great Rivers' – a fine title – but in their caution they had not given the man any salary or power. The second Inspector General was given some power and salary and he was able to do much for the country. His name, *Christiaan Brunings*, will be mentioned later. In 1798 a Central Authority of the Waters in the Netherlands was finally established under French influence.

It was this Central Waterstaat which planned our harbours to be fit for ships of all sizes, which devised new wide canals with large locks, and later ordered good modern roads and bridges to be made wherever needed. And the contractors carried out those works so efficiently that Holland became one great model of modern clean usefulness. While other countries were not even dreaming of national planning, Waterstaat could already study the main possibilities of development of the country as a whole. It could not only consider its physical and economic possibilities, but had acquired the power to carry out the schemes as soon as the money was provided. This power was exercised with care as soon as we had a *progressive Ruler* capable of taking the initiative.

In 150 years we produced three such Rulers, which does not seem much, but I doubt whether any other nation has produced more country-developing

---

[1] For instance the author Dr. Johan van Veen himself was awarded a golden medal for his thesis. This book ('Onderzoekingen in de Hoofden') was edited with financial assistance of 'Het Bataafsch Genootschap'. *S.*

Rulers in that space of time. The names are King William I, Thorbecke, and Lely.

We shall see the results achieved after Napoleon had had his way.

10. *A Kingdom for a Dredge!*

The time of Napoleon was a turning point in our history, since we invited the revolutionary French soldiers to come. Napoleon came in their wake and said: 'Ville assiégée est ville prise' and he proved the correctness of his theory for several years – but not for always.

Holland is *permanently* besieged by the sea. This enemy has been repelled again and again by human action and invention, but ultimately, if we frail human beings cease to be sufficiently alert, it will take its own again. The capture of this besieged 'town' of Holland by the sea may occur during some war, when bombs have broken the dikes and when our hands are tied too much to repair the gaps in a short time. For those gaps must be closed quickly or the lower half of the country will be lost beyond redemption. A task of at least the magnitude of the Zuiderzee works would be necessary to recover such a loss. Or shall we be driven away to the east – all of us – in order to make place for some alien race, as Hitler wanted to do in 1942? A highly-developed country, well situated, is worth stealing. For us eternal vigilance is vital.

The 18th century had been a century of decline in many ways. Every man and woman had heard the frivolous but ominous words of the Frenchman who said: 'Après nous le déluge' and the continent of Western Europe had indolently awaited the coming of that deluge. A general threat was hanging in the air like a thunderstorm approaching its outburst, slowly but steadily. When it had come and eventually passed, it left Holland a dilapidated country.

When the facts are weighed which caused the loss of hegemony after 1750, the blame is generally laid in the 'fatness' of the upper classes, but there also were natural odds against which we had but little redress; those of the sea and the rivers. The sea is slow and deceptive in its attack. It takes centuries to learn its wiles and it never relents.

I shall describe three of the besetting troubles of the 18th century. The first was superable, the second was not deadly but was insuperable, and the third was the most baffling of all.

The first trouble was the appearance of a *sea worm*. Presumably it had arrived with a ship from some far shore. It had a hideous form. The Dutch coast consists of loose or soft material and the weak spots of the coast had been protected against the scouring action of the sea by means of massive wooden shore defences. Once they had made them, people felt safe behind these strong and expensive works.

In the year 1730 the worm appeared and ate up the wood of the sea defences! This sent a wave of fear through the whole population. It was a national disaster. The land had become defenceless against its worst foe. It was useless to make new wooden protections. The Churches ordained public prayers for the removal of this 'terrible scourge of God'. Even a class of poetry was born which has become known as the 'worm-poetry'. Holland seemed lost because of a mere loathsome worm!

However, in due time, two clever men, *Pieter Straat* and *Pieter van der Deure*, invented the stone defence. Their names should not be forgotten, for their simple but good invention proved to be *the* solution; we still make these stone defences. The worm Limmoria is still with us and is eager enough to eat any wood standing in the sea except tropical hardwood, but we use stones now. However much we should like to use some home produced material to replace these stones, which have to be brought from remote lands, we have not succeeded in finding a good substitute, though concrete and baked clay go a long way. Asphalt may be a tribute.

The second trouble arose in the *Great Rivers*. We had made the Golden Hoop not only alongside the sea coast, but also along the banks of the Rhine and Maas and their effluents. The two rivers seemed to have submitted to the restraint of our dikes, but in the slow and grand ways of Nature, they had started an offensive. Patiently they had raised themselves above the land. The fluvial sands, brought from the interior, heightened the river bottoms bit by bit; now that the dikes were there, those sands could no longer spread over the country. The rivers became choked with their own sand; shipping became tedious, ice dams occurred every winter.

While building the river dikes we had started a competition between man and Nature. Man had heightened the dikes and Nature heightened the bottoms of the rivers. Then man had heightened the embankments again, and so on. But in the course of several centuries man had practically reached his limit, whereas Nature would go on endlessly bringing its sand from the hinterland.

So, whenever ice choked the rivers, and this was the case almost every winter, the dikes broke. When an ice dam had formed and the water rose dangerously, the church bells tolled in despair in every village, warning the people to flee quickly with their belongings and their cattle to the dike – the only high place in the low country. In times of such danger the whole population stood on the dikes. The men and women heightened them with sandbags, with wooden poles and boards, or with grass turfs. The dike master rode on horseback over his particular stretch of the dike encouraging the men and women, or calling them to some extra dangerous spot. Well-heads spouted behind the dikes when the water pressure forced a way underneath, or the water ran over the top when it found some low place, creating a rill in a few moments. The men rushed to

BUILDING IN MORASS AND WATER
by means of wooden piles. (Picture from about 1700).
Millions of pinetrees were driven into the ground by means of heavy falling weights. For every weight a group of about 30 men was used. Upon these piles the houses were built. In this picture a shipping lock is being made.

the rescue at such a spot, but often the dikes broke and then the waters thundered through the gaps once more.

How could this problem be solved? Some said: 'Make the dikes still stronger and higher', others: 'Don't fight against Nature any longer, demolish the dikes'. (These latter preferred the natural low floods to the catastrophic high ones.) A third category suggested abandonment of the least important regions to the floods, thus saving the best areas. It is a problem to be found in every land bordering the mouths of large rivers carrying sandy material.

This last advice of making spillways was considered the best. But what about the inhabitants of the regions which were to be abandoned to the floods? If our neighbours on the opposite shore had a weaker and lower dike than we had on our own side, the floods be for those neighbours and not for us! It has been this competition as to who could make the highest dike that had led to high dikes on either side. But who was to ordain now whose dike shoud be lowest? Was my land to be preferred to my neighbour's, or was he to be saved and I to be drowned? To decide this was no easy task for any Government.

81

COASTAL DEFENCE BY MEANS OF STONE GROYNES AND A STRONG DIKE

A ship has stranded. Behind the dike a lake was formed by removing the substance needed for the dike. *(Photo K.L.M.)*

Spillways were made, however, through which the rivers spent their freshets and ice – but not their sand! The fluvial sands remained inside the embankments; the rivers were unable to carry this sand either into the sea or into the traverses. Making spillways in a well-populated and fertile land is a rather poor solution, but even then the cause of all the trouble – the sand coming from the interior – is not removed so long as it is not carried into the sea. And this was the one thing we could not make the rivers do. Even in 1850 we had not solved that awkward problem.

Shipping along the Rhine suffered so much because of this sand that after the Napoleonic frenzy, when King William of Orange wanted to inspect this most important river, his yacht of only 3 feet draught grounded. We thought it a shame at the time that our main river should be in such a bad state, but we could do nothing about it. It was the steam dredge which brought the solution after 1860. All the sand which had come down from the interior had to be dredged away.

The third trouble was the worst; it crippled us as a trading nation.

There are signs that our leading classes of the 18th century were not so far-

82

sighted as their ancestors had been. But there was also a natural cause which cannot be disregarded: *the lack of depth* in our harbours.

To have ports cut off from the sea, a whole nation feeling lamed because of some odd trick of Nature which cannot possibly be mastered, is not an uncommon experience in history. Many are the towns which have seen their harbours silted and become doomed to insignificance.

We had to face this disaster. Since 1600, when it had sufficiently freed itself from oppression, Holland had become the greatest shipping nation in the world and now the fall was great. To gain an idea of Dutch trade some figures will be given. Even few Dutchmen know these figures. They are mainly to be found in English books, because the English studied the cause of our success carefully in those days, in order to use the same successful methods.

In 1588 the Spanish 'Invincible Armada', built to conquer England, had 132 ships. It was considered a mighty fleet; some ships were of more than a thousand tons. The English fleet sent to beat off the Armada had 197 ships of about 300 or 400 tons each, many of them taken from the merchant fleet. In 1600, that is 12 years later, Holland possessed about 20,000 sea-going vessels, which was more than the whole of the rest of Europe could boast. The Dutch fleet in the Baltic trade outnumbered the English Baltic trading fleet by ten to one.

In 1650 Holland had 14,000 to 15,000 ships, which was about 60% of the European trading fleet. They were larger than those of the year 1600. A French traveller who visited Amsterdam in those days said that as many as 3000 or 4000 ships might be found lying in port at one time. Twice a year fleets of 400 to 500 ships each arrived from the Baltic or from the Mediterranean. Other fleets arrived from time to time from the Indies and other newly-discovered countries.

The Englishman *Child*, who has been called the father of the modern science of political economy, described the Dutch trade in 1665 as 'the envy of the present and the wonder of all future generations'. The Ambassador *Sir William Temple*, who resided in Holland for a long time and who knew the country very well, wrote a book about Holland in 1673, explaining the Dutch prosperity. It immediately became a best seller. This prosperity, due to the 'industry and parsimony' of the Dutch, was not a mere accident, Temple said, 'but the result of a great concurrence of circumstances, a long course of time, force of Orders and Methods, which never before met in the World to such a degree or with so prodigious a Success, and perhaps never will again'.

In 1700 when the other European nations had already reached a certain advancement, having learned the Dutch 'methods and order', Holland still had roughly twice the amount of shipping possessed by the British and ten times more than the French. In that year one-third of the students of Leyden University were English or Scots, whilst many other Britons studied at other Dutch universities. The Bourse of Amsterdam was still 'the hub of a gigantic network of

DIFFERENCES IN THE MAIN FRAME
1750

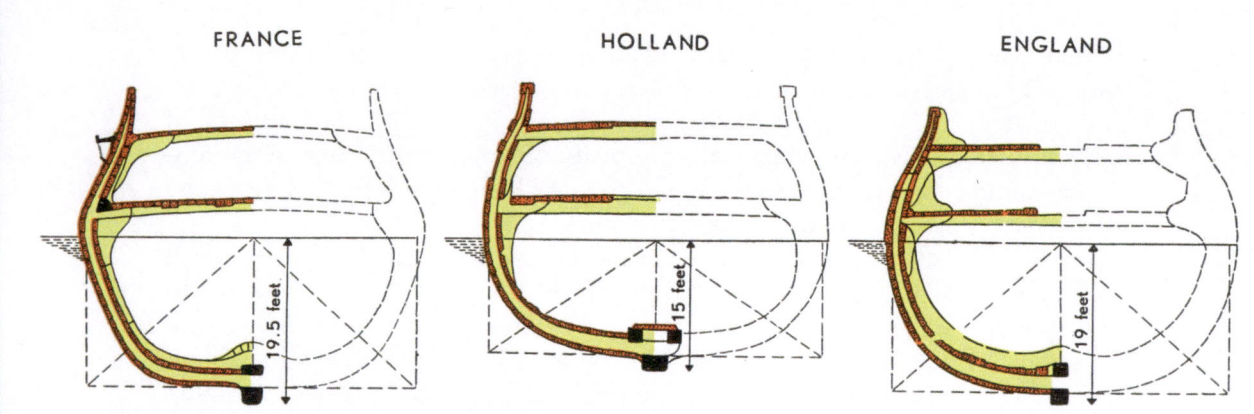

FRANCE   HOLLAND   ENGLAND

Main frames of ships in 1750, showing how the Dutch ships could not compete any longer because of lack of draught, owing to lack of depth in their harbours and inlets (after E. van Konijnenburg). (Depths in Amsterdam feet.)

world trade, shipping and finance', where the European goods were sent to be sold and transported. Amsterdam was the biggest trading town in Europe.

Yet, the sea entrance of this biggest trading centre of the world had never had a greater depth than 3¾ metres, which is slightly more than 12 feet! As far back as 1400, a bar at the entrance of the port of Amsterdam was mentioned. In 1544 Charles V, Emperor of Spain and of the Low Countries, ordered that ships should have a draught not exceeding 12 feet. But in the 17th and 18th centuries this depth in the entrance of Amsterdam harbour decreased to 9 feet below low-water, or 10 feet below high-water. In 1859 the bar of Amsterdam had still a maximum depth of 10 feet at high-water, but the ships could have a slightly greater draught if they were pulled through a layer of mud.

Another drawback was the silting up of the harbour of Amsterdam itself. The handicap provided by Nature was overwhelming. The ships of the competing nations could be given a better shape; we could not give sufficient draught to ours. For the same size of ship, with a width of 40 feet the following heights inside the ships between bottom and deck may be compared:

British ship . . .   18½ Amsterdam feet
French ship . . .   20¾   ,,   ,,
Dutch ship . . .   16   ,,   ,,

From this it may be seen that the Dutch vessels acquired a most inefficient shape. This was about the year 1750, when it had been found that larger and deeper ships meant higher speed, more carrying capacity and a much higher earning power.

Neither Rotterdam nor any other port in central Holland had more depth in its sea entrance than Amsterdam. The Dutch ships had to be made broad and shallow, and therefore clumsy. They became too slow. It had been in Holland

84

that the science of shipbuilding had created a new type of ship which was used all over Europe. It was a development embodying the advantages of the two ancient European types: the Norse one – built for speed, not for trade, because the Vikings were pirates – and the Mediterranean type. By combining the speed of the Viking ship with the roominess of the southern type a good merchant ship had been obtained. As early as 1650 the Dutch wharves produced 2000 sea-going ships every year, 'having not one timber tree growing in their own country, nor homebred commodities to load 100 ships, and yet they have 20,000 ships and vessels and all employed', remarked Raleigh in 1603.

This great advance in the art of shipbuilding had led to a point where we ourselves could not follow, because our sea entrances lacked depth. The more ships increased in size, the more difficult it became for the Dutch Navy to protect Dutch interests in the world; and in peace time, trade suffered because our ships were too small and clumsy.

In 1671 *Nicolaas Witsen* wrote a famous book about shipbuilding in which he said that frugality and cleanliness were worth more than the form of the ship. This thesis may have been invented to put some heart into our skippers, but nevertheless no frugality or cleanliness could prevent London from becoming the trade centre instead of Amsterdam. Thus the art of building fine great ships went to France in the course of 1750 to 1790 and then to England. The Swedes had said of the Russian fleet of Czar Peter the Great: 'that they could not see anything Russian on it but the flags and that, if thus fleet had to fight against the Dutch, such had to be done under Dutch admirals commanding Dutch sailors and with Dutch gunpowder shot from Dutch cannons.' But those days were gone.

The sea had beaten us with its sands, as the Great Rivers had done. The hegemony of the Dutch which had reached such grand heights for a period of a century and a half passed from Holland to England.

Yet, despite the severe drawback of the shallow sea entrances, it was not so much a fall in total tonnage and trade, but the check in development which oppressed. The maximum number of Dutch ships passing the toll of the Sont was in 1609 –19, when 32,903 of them were counted. In the so-called bad years 1770–79 23.030 Dutch vessels were counted, but the average tonnage had increased by then.

It was Napoleon who finally came very near to extinguishing the whole of the Dutch merchant fleet, when he wanted to conquer England by forbidding sea trade of every kind. Holland could not live without the sea.

The narrow circle also narrows the mind! Over the 'narrow circle' of the Netherlands now dropped a curtain on which was written: 'We are so small'.

We felt conquered, humiliated and, being cut off from the sea, had contracted the smallness-complex of which Schiller says that there is only one way out.

Nothing other than a noble aim
Up from its depths can stir humanity.
The narrow circle narrows too the mind,
And man grows greater as his ends are great.

A noble aim with un-narrow limits was needed, and *King William I* of Orange, raised in the victorious and soaring England of his days, was the first to detect and shape this aim. He opened the doors to the sea, and the sea is never small. Its greatness makes for greatness.

II. *The Spade once more*

The sad inheritance of Waterstaat in 1815 was a memory of past glories, a couple of dead cities as museum pieces, and a collection of silted harbours and choked rivers. But the deluge had had its way and the sky was no longer so threatening.

When King William of Orange set foot on shore after the Napoleonic wars, he was eager to lead the nation to new prosperity. Waterstaat had to solve the physical problems. The river floods had to cease, King William told the engineers. The rivers had to be made deeper, so that the ships from the hinterland might come and go. Moreover, the King said, the ports of Amsterdam and Rotterdam had to be made accessible to all kinds of modern sea-going vessels. How could this be done? Seaworthy dredgers had not yet been invented. True, as every Dutch schoolboy knows, *grass grew in the streets* of the towns of Holland, and something positive had to be done to make a living. But how?

King William was an energetic and resourceful man. It is said that he took a *thick pencil* and with it he drew a rough line on the map from Amsterdam to a place in the North of the country, called Den Helder. He wanted a canal to be made there for sea-going ships. It had to become the widest and best canal in the world.

He drew another line straight across the island of Voorne (an island West of Rotterdam) to connect Rotterdam with the port of Hellevoetsluis, where the sea allowed for some depth.

In this way the new King of the ancient and famous House of Orange proved his initiative. He even went so far as to finance the works personally to a large extent. They were excellent ideas to help Amsterdam and Rotterdam – and therefore the whole country – because we could make those canals with the spade, the wheelbarrow, the horse-drawn cart and the wooden horse-driven mudmill. We could also make locks large enough for the biggest ships in the world.

Moreover, in the years before the French revolution, there had been the Inspector of Waterstaat, *Brunings* by name, who had made a deep and wonder-

86

F. W. CONRAD (1769–1808)

Conrad was Brunings's successor. They both lie in the same grave in the Groote Kerk at Haarlem, a place of great honour. On this grave is chiselled: "Counsel and Protector of the Netherlands against the Fury of the Sea and the Storms".

CHRISTIAAN BRUNINGS (1736–1805)

Brunings was the first Inspector General of the Rijkswaterstaat, the "Head which thinks, and cares, and watches while the floods rise high, warding off the threat against Bato's low estate". (Bato is the supposed first Batavian or Dutchman.)

ful harbour at Den Helder in 1782. It could hardly be said that he had made it or that the Dutch had made it, for he hadflet the ebb currents do the work. He had only made a simple dam on the sand lats and had thus obtained between the dam and the shore a funnel in which the ebb was forced to scour a depth. This depth was sufficient for any ship. It was considered a masterpiece of applied hydraulic intelligence. Brunings' harbour was extended in 1954, but it has been our naval base from the beginning of its existence.

The new canal was dug to this fine deep harbour of Brunings. Its dimensions were unprecedented; its length was 49 miles, approximately the length of the Panama Canal or 60% of the length of the artificial part of the Suez Canal. Its depth was 16½ feet and sea-going ships of the largest size could easily pass. It took from 1819 to 1825 to construct. Taking into account the fact that it was made 40 years before the Suez Canal and therefore without steam-dredgers, it was a great work. In those days the new locks of 46 feet width must have been the largest in the world. The canal for Rotterdam was likewise rapidly dug to the port of Hellevoetsluis and the width and depth of this canal and its locks were of the same unprecedented size.

King William had made a good start. The new era, however, had no longer clay feet to walk with, but wings. The size of sea-going ships increased so quickly that these, the widest and best equipped canals of the world, soon became too narrow. It was a pity, but it was not King Williams' fault. When we visit his canals now, it seems unbelievable that those were the finest canals of their time and we wonder what a sight it must have been to watch stately Indiamen being towed through them by horses. Conditions changed greatly in a short time. It was the steam dredges that made the huge modern shipping possible.

12. *The Birth of the Dredge*

The decline of the 'richest and most urbane nation of the world' according to the English, the fall of 'le nid de richesse', according to the French, at the end of the 18th century had not occurred without a struggle. For four centuries inventors had been at work to ward off the threatening calamity of silted harbours. A really ingenious invention was needed.

The Dutch coast is a changeable one; the sands and silts are easily displaced by the currents and the waves. In a land like this, where physical conditions are unstable, ports of one age may be found far inland in the next. In some towns we can see till the Water Gate, where the ships used to moor, but instead of ships and a harbour we find cows and sheep grazing. Nature, in a statu-nascendi country shifts and removes huge quantities of sandy material slowly but incessantly, and man's wit must devise means to defend himself lest his works be buried under those materials.

One of the first successful instruments used for removing silt was the *Krabbelaar* or Scratcher. Already in 1435 the town of Middelburg (Zeeland) used such a Krabbelaar and it must have been invented earlier. Other names for it were the 'Water-harrow' or 'Water-plough', but people called it 'the Mole', because of the way it routed up the mud. The Krabbelaar was used at ebb-tide. Its harrow or plough loosened the bottom material of the harbour entrance while the ebb was flowing. The shiplike instrument was moved by wind, having sails for this purpose. It was also moved by the ebb itself, as it had wings under water which could be spread to catch the currents. As late as 1800 there were still such Krabbelaars in use in some parts of Holland.

Another instrument was the *ship's-camel*, invented in 1688 especially for the

THE 'KRABBELAAR' (1435)

Iron harrows underneath the ship scratched the bottom of a channel during ebb tide. Sidewings and sails helped to move the Krabbelaar, 'Scratcher' or 'Mole.'

bar in the entrance of the port of Amsterdam. The idea was very simple; two halves of a ship were fastened on either side of a merchant vessel. By pumping water out of those 'camels', the vessel could be raised partly out of the water. The camels were unwieldy wooden affairs. The pumping had to be done by manpower and there were wooden barrels inside the camels which were heavy in themselves. These ship-raising 'Camels' could not be used where there was a possibility of heavy seas, but luckily the Zuiderzee near Amsterdam was mostly calm. They were considered a nuisance but they could not be dispensed with. They could raise a ship about 7 or 8 feet.

Then there were the *tidal reservoirs*, filled by the tide at High Water and emptied at Low Water, which could be used to create a scour. This method produced more results when a Krabbelaar was used in conjunction with the emptying of the reservoir. At Amsterdam such reservoirs could not be made, because the tidal rise in the Zuiderzee was no more than one foot. Also at a

THE FIRST MUD-MILL (1600)

Treadmill, used to deepen the harbour of Amsterdam. The silt was shoved up a gutter by means of boards on a chain.

place such as Rotterdam this system was useless, because the bar was out at sea in the mouth of some wide river. A few minor ports could be helped somewhat by means of tidal reservoirs, but they were generally unsatisfactory. The best invention of all was the famous *Amsterdam Mud-mill*. It must have been invented before 1600 and from it our modern bucket-dredgers were derived.[1] The idea was already much the same, but every part was made of wood – even the gears and the buckets, or boards, with which the mud was shoved up a wooden gutter. The power by which this first dredge was driven was provided by men operating a tread wheel. In 1620, however, there was a horse-driven mud-mill in use. The horses trotted in a circle on deck, while other horses were kept in reserve in a stable on board.

The horse-driven mud-mill proved to be *the* invention that was needed,

[1] Thanks to the recent investigations of Mr. G. Doorman (1952) we now know the name of the inventor being Cornelis Dircksz. Muys. He gives the following description. 'In general terms Muys's invention was a dredger with an inclined chain, running over an upper and lower tumbler, taking up the mud near the bottom and throwing it out above, its depth in the water being adjustable by a swinging movement of the chain around the axis of the top tumbler. Such a definition would cover all modern bucket-dredgers.' *S.*

90

AMSTERDAM MUD-MILL (1650)

Horses provided the driving power (left). The stable is at the right. The windlass served to lower the ladder, on which the mud was shoved up. The mud-mill was in use from before 1600 until 1860.

even though the wooden gearing could not carry much strain. This gearing resembled that of the windmills or of the ancient churn mills which may still be found on some farms. From the beginning it proved to be a useful machine. It held its reputation far into the 19th century and was able to compete with the first steam dredges. Most European ports ordered mud-mills to be made at Amsterdam and the construction was considered so good that there were hardly any changes in the original design. An excellent book published at Amsterdam in 1734 describes the highest engineering art of the time – '*The Great Complete Book of Mills*'. It shows what mankind could make in this era of wood mechanics.

Amsterdam levied mud taxes to cover the expenses of dredging. Between the years 1778 and 1793 the tax totalled 1,598,000 guilders, quite a large amount for those days. There were generally 5 or 6 mud-mills in regular use at Amsterdam, each mill having 5 horses on board, of which 2 were pulling and 3 resting. Every day 400 tons of mud could be dredged with one mill and depths of 10 to 15 feet could be obtained with them. But the bar in the Zuiderzee could not be dredged away with these instruments; nor could the mud-mills be used in sea inlets where the waves would destroy the clumsy structure in a few moments.

Towards the end of the 18th century the English invented the steam engine and after a while they produced the first steam dredge. Amsterdam bought

91

one, but found that it could not compete with the old mud-mill. This was so even in 1835, according to the Dutch-German expert *Van Ronzelen*, the man who made the port of Bremerhaven. After having come to Holland to study hydraulic engineering, he reported to the Prussian Government that there were 16 horse-driven mud-mills at Amsterdam and one steam dredge, but the latter was not used. The mud-mills were able to make and maintain a depth of 15 feet easily, he wrote, and one had been made which could even make a depth of 21 feet. With one large modern horse-driven mud-mill 540 tons could be dredged in 3 hours! Could the iron steam dredge produce more? It could not, and, moreover, it was 7 or 8 times more expensive in initial cost. So, why buy English or French steam dredges, when the old Amsterdam mud-mills were so much cheaper and better? Van Ronzelen's advice was therefore to order and use the Dutch-driven mills. This advice was based not on conservatism, but on the best of all reasons – economics.

Towards the end of the 19th century there was still a mud-mill constantly in use in the canal from Amsterdam to Den Helder. It proudly bore the name of that canal, 'Great North-Holland Canal'. It dredged 400 tons a day at a depth of 22 feet. An advanced engineer, long accustomed to iron machinery, saw it there and wrote with humour of this last of a hard-working race of mud-mills. He forgot that these mills had stood in the heat of the fight for Holland's hegemony and that they had kept off all dredging competitors for more than two and a half centuries. He joked about its archaic appearance and called it 'Noah's Ark', but failed to see that the steam-driven bucket dredge was the direct descendant of this old faithful mud-mill.

There were some other inventions, for example the grab, the shovel and the dredging wheel. Of these the grab and the shovel have had important developments. The ship's-camels developed into modern floating docks.

The first use of a real bucket dredge was mentioned in 1623, when the river Maas was deepened with it in front of the town of Heusden in Brabant. The buckets were made of copper and were 'made to rotate' on the ship by means of manpower. 'Much sand' was dredged and 'a fine depth was obtained thereby'. The sand fell into another ship lying behind the dredge. It was the combination of this copper bucket dredge and the mud-mill which produced our modern type of dredge.

13. *The Rotterdam Waterway*

The fault of the ancient mud-mill was that it was not fit for sea-work. So long as the iron dredge was equally unfit for work at sea, little progress could be made. In the first half of the 19th century we had to wait for powerful steam-dredges having greater capacity, which would allow our sea entrances to be

Originally Rotterdam could be reached by the inlet north of Voorne, which was dammed off in 1950. Afterwards by one of the waterways I, II or III, allowing uncertain depths of about 15 feet. In 1829 the Canal through Voorne was dug, which soon afterwards became too small. In 1872 the Waterway (IV) was made to the Hook of Holland. All the docks of Rotterdam are open tidal basins.

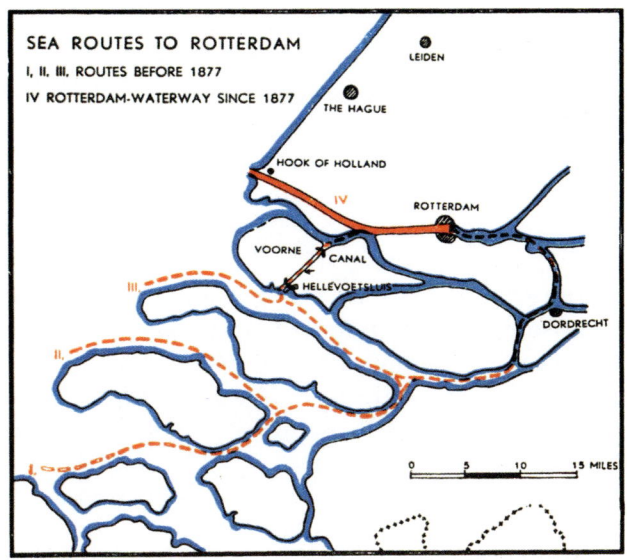

SEA ROUTES TO ROTTERDAM

I, II, III, ROUTES BEFORE 1877

IV ROTTERDAM-WATERWAY SINCE 1877

LEIDEN
THE HAGUE
HOOK OF HOLLAND
ROTTERDAM
VOORNE
CANAL
HELLEVOETSLUIS
DORDRECHT

0   5   10   15 MILES

deepened. We were also waiting for the man of authority who would say that he wanted those better sea entrances, a man of sufficient force to back more-advanced planning. This man was *Prime Minister Thorbecke*, the staunch sponsor of the Rotterdam Waterway.

King William, about 1825, had made sea entrances to the ports of Amsterdam and Rotterdam which can be regarded as excellent for those days. It had not been his fault that the size of ships increased so rapidly that his splendid canals, the biggest in the world, soon became too small. After three decades had passed, our entrances from the sea were again insufficient to allow the biggest ships to call at our ports.

'*Our coast must be made accessible for the big shipping of our times*', Thorbecke said to the applauding House. This was in the year 1862. Thorbecke not only wanted good harbours for his own country, but he saw the possibility of opening up the hinterland. He compared a harbour with a 'funnel, through which world trade can come into contact with the half of Europe, with all the land that lies behind us, . . . The wider the funnel, the more can go through it'.

The plan which Thorbecke had in mind when he spoke these words was one of long standing. It was to give one of the main branches of the Rhine a new mouth by cutting through the dunes at the Hook of Holland, so as to provide a new deep open river by which Rotterdam could be reached from the sea.

The originator of this idea was the engineer *Kruik*, or Cruquius, the man who had wanted to study everything appertaining to Waterstaat as early as 1726. Now, an engineer of Waterstaat had revived that plan and improved it. His name was *Pieter Caland*. His idea was to make the currents do the main

93

work. First we would have to make a small initial cut through the dunes into the sea, secondly we would close the existing wide but shallow mouth of the old river and then the streams would, he hoped, widen the initial trench and make a new deep river.

Since the invention of the steam-dredge the risk and romance, as well as the art of hydraulic engineering, may have partly vanished. But it had not yet vanished in 1860. *De Lesseps* had just started his Suez Canal (1859–1867) and neither *Lord Palmerston*, the diplomat, nor *Robert Stephenson*, the engineer, had any faith in the success of that work. Yet the dredging of the Suez Canal, as seen from a hydraulic standpoint, could hardly be called a risky work.

Caland's idea was risky. He wanted to induce Nature to make a new river mouth, deep enough for the deepest ships in existence. He must have had in mind the success of Brunings who had done more or less the same thing when he made the harbour of Den Helder in 1782 by scouring. At the time we were proud to see that human ingenuity could command the power of Nature, usually so opposed to our purposes. But would the streams actually do the scouring work at the Hook of Holland? Would Rotterdam become a mighty harbour?

Thorbecke said he had no choice. He continued his speech in the House by saying: 'Often in this discussion I have been quoted as saying that this work of making a Waterway for Rotterdam is a risky work. I still say so. It is not a recommendation, but I will not go further than what I consider to be just and true. I still say that it is a risky matter, but it is a risk we must take.'

The climax of his speech sank home when he said: 'It is a case like that of an uneven struggle taken up for freedom and independence. *If we remain what we are now we shall be outdone and shall be lost.* It therefore appears to me that it is our inescapable duty to grasp the hand that can save us.'

So the risky work was started. The initial ditch was made a little less than 7 feet below Low Water, and its width at the bottom was no more than 30 feet. Then the existing river mouth was dammed and the scouring of the new tidal river could begin.

It did scour slightly, but not nearly to the extent Caland had anticipated. Even as late as 1873 the depth was still no more than 10 feet. Caland's theory had evidently failed!

Peter Caland fell into disgrace. His father Abraham, one of the best dike-engineers Zeeland produced, wrote to him in 1868: 'Come Peter, come to your old father and bring some papers and maps with you, that we may talk things over and find courage to persevere. Hundreds of engineers have suffered adversity with extraordinary works, behave like a man. Hold fast to your science.' But Peter Caland's star fell.

The country became impatient. Rotterdam had now waited several decades for a good sea entrance, and it had not obtained it. 'Our coast must be made

ROTTERDAM WATERWAY

The sea is in the background. The Waterway was largely made by dredging. Annual shipping tonnage before the 1940/50 war was about 24,000,000 metric tons. Depth at HW. is about 40 ft., but it can easily be made greater.   *(Photo K.L.M.)*

accessible to the big shipping of our times', were the words, or variations thereof, which had been heard since 1700. But for more than 150 years this sandy coast had not been accessible for big shipping. Impatience demanded further action, if not from Nature, then from men.

It was now proposed to make locks in the Rotterdam Waterway at the Hook of Holland. It was evident, however, that an open river was to be preferred.

How to obtain depths at the sea side of the locks? Again there was a dead-lock in the discussions.

But new methods had arrived. Dredging in the open sea had at last become possible and so it was decided to dredge, and to dredge thoroughly, regardless of cost. Waterstaat now resolved '*to break for ever and ever with the system of scouring by the currents in the Rotterdam Waterway*'. We can perceive some satisfaction and pride in this sentence. It is the feeling of having at last obtained mastery over one of the most baffling problems which Nature had presented in the course of many centuries. Had faith in the benevolent action of Nature proved to be a failure once more? If so, collaboration with Nature had to be discontinued. Nature or no Nature, man wanted a deep and wide river quickly and would

95

not wait any longer. He now made it with the powerful new tool, the *steam-dredge*. In the same year, 15 dredges of the biggest type were set to work and they quickly deepened the Waterway to Rotterdam satisfactorily, making a wide and very good river.

This mastery over Nature was a milestone in the history of the Dutch nation. Gone were the days when Holland was doomed to a sad decline because of the shallowness of its waters. Nature was conquered once more! We no longer were forced to suffer poverty, and lose trade, nor stand impotently and baffled on the quays. The gates to our country were open again.

There was no longer any excuse for lagging behind competitors, and Rotterdam promptly became one of the greatest ports of the world.

14. *The Amsterdam Waterway*

We have left Amsterdam with its canal to Den Helder, which in its day was the biggest canal of the world, but soon to be overtaken by the increase in the size of ships.

King William's first idea had been to cut a new canal straight through the dunes west of Amsterdam, instead of going to Den Helder. The engineers had said, however, that making a new canal direct to the North Sea would be impractical. To cut the broad ridge of high sand dunes might be possible, but to make a harbour on the sandy shore, where no depth at all was available, was considered quite unfeasible for the time being. Of course it was, so William gave in.

This cut through the dunes was nevertheless a good and also an old idea. As early as 1634 a certain *Jan Dou* had advanced it, in order to obtain an outlet for the draining of the low country behind the dunes. In 1772 the famous engineer *Brunings* wanted this canal through the dunes for the double purpose of giving Amsterdam a better sea entrance to the North Sea direct, and at the same time securing lower tides for more effective drainage of the low land of Central Holland. Again in 1824, when the King made his choice of the Canal to Den Helder one well-known Waterstaat engineer had advised the direct route to the sea.

Once the success of the Great North-Holland Canal to Den Helder started to wane, other voices went up to back the original idea. According to enthusiastic supporters the opening of the dunes west of Amsterdam might even provide an open sea inlet without locks; the tidal currents might be able to create the depths and bring the ships in a very short time to the port of Amsterdam. Amsterdam, originally situated on the shores of the shallow Zuiderzee, would become a port directly on the North Sea!

It might be asked here why such a great seaport as Amsterdam could ever have developed on such an insignificant bay as the Zuiderzee. Amsterdam's hinterland originally consisted of nothing but morasses and wildernesses. Its soil was most abominable – the houses having to be built on piles of about 30 feet length. A waterway from Amsterdam to the Rhine hardly existed, as the

After the Zuiderzee was formed, about 1300, Amsterdam came into being as a port for the newly reclaimed peat marshes of central Holland. Route I had a depth of 11 or 12 feet and was used until 1824. Route II was a new canal used from 1824–1876. Route III (1876) was made fit for the biggest sea-going ships. At IJmuiden a lock has been made through which a ship of 150,000 tons, having a draught of 50 feet, can pass.

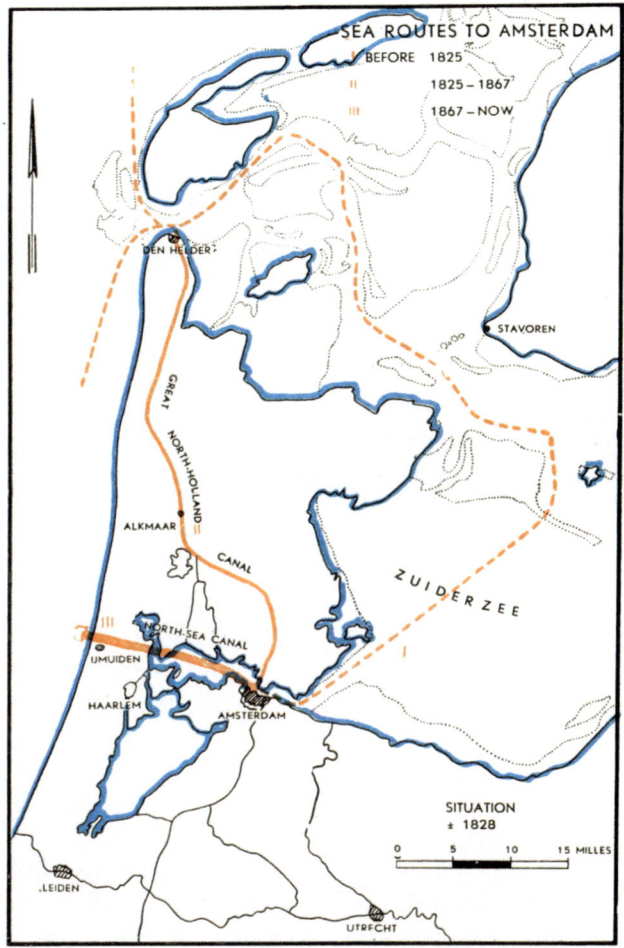

old branch of the Rhine, called the Vecht, had been shut with a dam. The ancient town of Kampen on the other side of the Zuiderzee was the port on the northern mouth of the Rhine and so was the old town of Dordrecht on the mouth in the west. There was no obvious reason why Amsterdam should become a great trading centre.

It would have been understandable if Rotterdam had been chosen by fate to be the leading town, because Rotterdam was situated at the mouth of the Rhine.

97

The largest lock in the world (IJmuiden). A ship twice the size of the Queen Elizabeth could pass through this lock. The crowd on the quay gives an idea of its size. *(Photo K.L.M.)*

We should take into consideration, however, that the powerful Hanseatic League had its centre at Hamburg and Lubeck, northeast of Holland, and that the Hanse ships could come to Amsterdam more easily than they could reach Dordrecht or Rotterdam. Also the effect of the shallowness of the Zuiderzee was not felt so much in those early days of shipping. The main reason for the rise of Amsterdam was, however, that the people living near the dam in the small river Amstel, were sailors and traders. The soft, soaked soil did not invite cultivation, but with their small ships they attracted trade much in the same way as squirrels accumulate acorns. Beer from Hamburg was one of the main trades, for apparently the pioneers in the immense swamps south of Amsterdam were thirsty and the swamp water unhealthy.

It was shipping which was the main cause of the greatness, not only of Amsterdam but of the Low Countries as a whole. Shipping – not trade – was the primary thing. The skippers created the trade, or invented it by finding new ways to launch it. The Dutch skippers, the Amsterdammers in the first place, mastered the Hanse more and more, and in 1440 fought a war against this League, in which the hegemony passed into our hands.

If we were to be once more confronted with the task of providing a Water-

98

way from Amsterdam to the North Sea, we might perhaps choose an open canal without locks as proposed by some of the best engineers of the 19th century. But in the year 1852 there were no good sea dredges capable of providing sufficient depth, and the engineers of those days had inadequate knowledge of tidal phenomena. They were afraid that a wide sea inlet might be formed which would become too powerful to control, and so in 1852 the leading engineers pronounced the open canal to be an 'inexcusable imprudence'.

Who will blame them for choosing the safe way? Canals shut with locks are safe. Locks are a nuisance, but the alternative was making a longer coast, making more high dikes along the new coast, inviting the salt water into the country, and severing road connections with the north.

In 1867 this direct 'North Sea Canal' was finished; about the same time as the Suez Canal was opened. Its cross sections were about the same as those of the Suez, the initial width at the bottom of the North Sea Canal being 82 feet and the depth 30 feet. At the moment they are: bottom 230 feet, depth 38 feet. In the future these dimensions will be increased to 300 × 47 feet. A village sprang up at the sea end of the canal, called IJmuiden. It is now becoming a large town.

Comparing this work with the Rotterdam Waterway we find different principles. At Amsterdam we preferred the canal closed with locks, at Rotterdam we chose the risky way of making a new river with the aid of the scouring action of the tides. One of the results was that the Amsterdam Waterway was finished earlier than the Rotterdam Waterway.

It is still difficult to say whether we chose rightly or not. Rotterdam with its tremendous transport of goods could hardly have locks. In the Rotterdam Waterway ships are often coming and going in double file and at regular distances, so that every few minutes two ships are passing, one going up-river, the other going down. Hardly any system of locks could be devised to cope with so many ships in such a short time. At Amsterdam, however, the shipping is not so frequent as at Rotterdam, and the silting of a canal is less than the silting in an open tidal channel. Yet, a lock is also expensive.

With the growth of ships the size of the locks at IJmuiden had to be increased proportionately. Three successive locks were built, which give an idea of the increase in the size of ships. They are:

	Width	Depth	Length	Proportion of contents
in 1867 largest lock:	60 ×	26 ×	392 feet,	1
in 1896 new lock:	82 ×	33 ×	750 feet,	3.3
in 1930 latest lock:	165 ×	50 ×	1320 feet.	17.7

These depths are taken below mean sea level. The latest lock is still the biggest lock in the world. The very largest ships can easily be handled there, because

IJMUIDEN LOCK

The groove shown here, into which one of the doors slides, has been pumped dry for repairs after the Germans had tried to destroy the whole lock. Note the size of the two men on the floor and especially the man standing at the end of the groove on the other side of the lock-chamber.

the dimensions of the Queen Elizabeth are only 118 × 36 × 1029 feet. When in due course ships of 150,000 tons are built, both Amsterdam and Rotterdam will be able to receive them.

The reason why we made such a large lock was partly because we had learned in the past that locks and canals for seagoing vessels soon may become too small. Also we wanted to be on the safe side for fear of losing trade again. The whole world was to know that any ship any nation may build is welcome in our harbours. We have learned our lesson thoroughly and intend to avoid the horrible situation we experienced from 1700 to 1860. Never again do we want to fall into that quagmire.

The sea-approaches to Holland are no longer shallow, nor the hinterland inaccessible. We owe this to the dredges. Perhaps some later generation will discover some beauty in a dredge and put it on a pedestal as a monument. It surely deserves it, for the dredges have given us deep water and prosperity.

Another beneficial factor was that some statesman or engineer has always jumped at every possibility which promised a slight chance of success. We do not blame Pieter Caland for having dared to try to gain more depth for Rotterdam by using the currents before good dredges were available. On the contrary we applaud him, as well as King William, Thorbecke and so many other influential people for having made such efforts to make our coast accessible again. They were never unready, but made courageous use both of their intelligence and of the latest inventions.

In the course of time more harbours were deepened, for instance Harlingen, Delfzijl, Vlissingen (Flushing), Schiedam, Vlaardingen and Dordrecht. Our gates are wide open now, inviting trade. The inland waterways, too, are in such condition that coasters of the normal type can reach every inland town.

15. *Improvement of the Rivers*

We have called the Rhine and Maas our 'second trouble' and explained that these rivers brought their sands to Holland but did not carry them out into the sea, so that in consequence they became choked.

At the Congress of Vienna in 1815 it was resolved that the border states of the Rhine should improve that river. It is fairly easy to alter the natural chaotic conditions of rivers in the middle part of their courses. The cure is to cut off some meanders, to give a regular width to the river and to make smooth banks as well as streamlined curves. But this cure is ineffective for the stretches near the sea, where the surface slopes are small and where there are no meanders to be cut off. Quite near the sea there is the 'devil in the water', as some sailors call this phenomenon.

The scientific name for this 'devil' is '*underflood*'. It is the influence of the salt

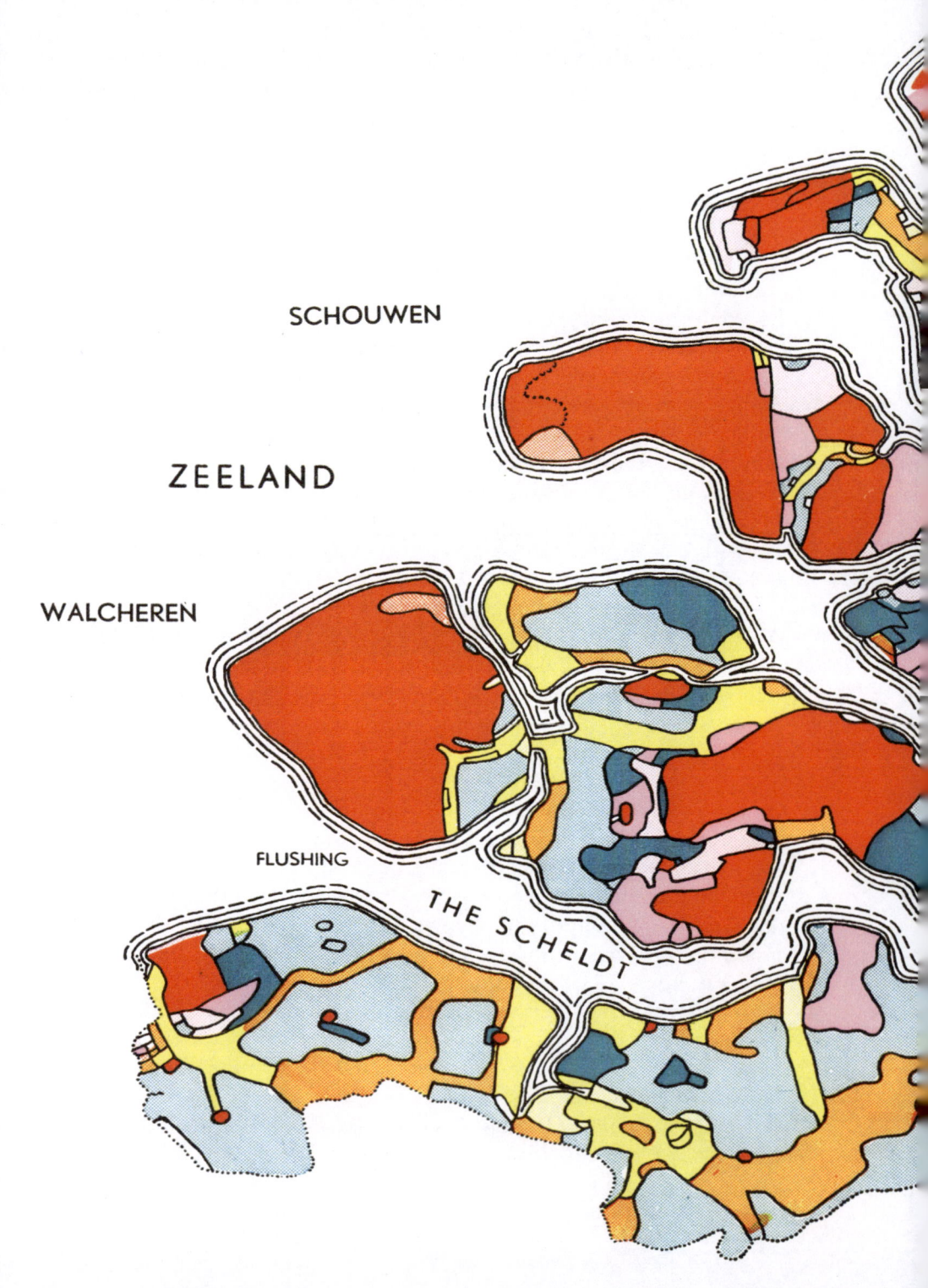

POLDERS GAINED
IN THE SOUTH WEST OF THE NETHERLANDS

HOOK OF HO...
WATERWAY

SCHOUWEN

ZEELAND

WALCHEREN

FLUSHING

THE SCHELDT

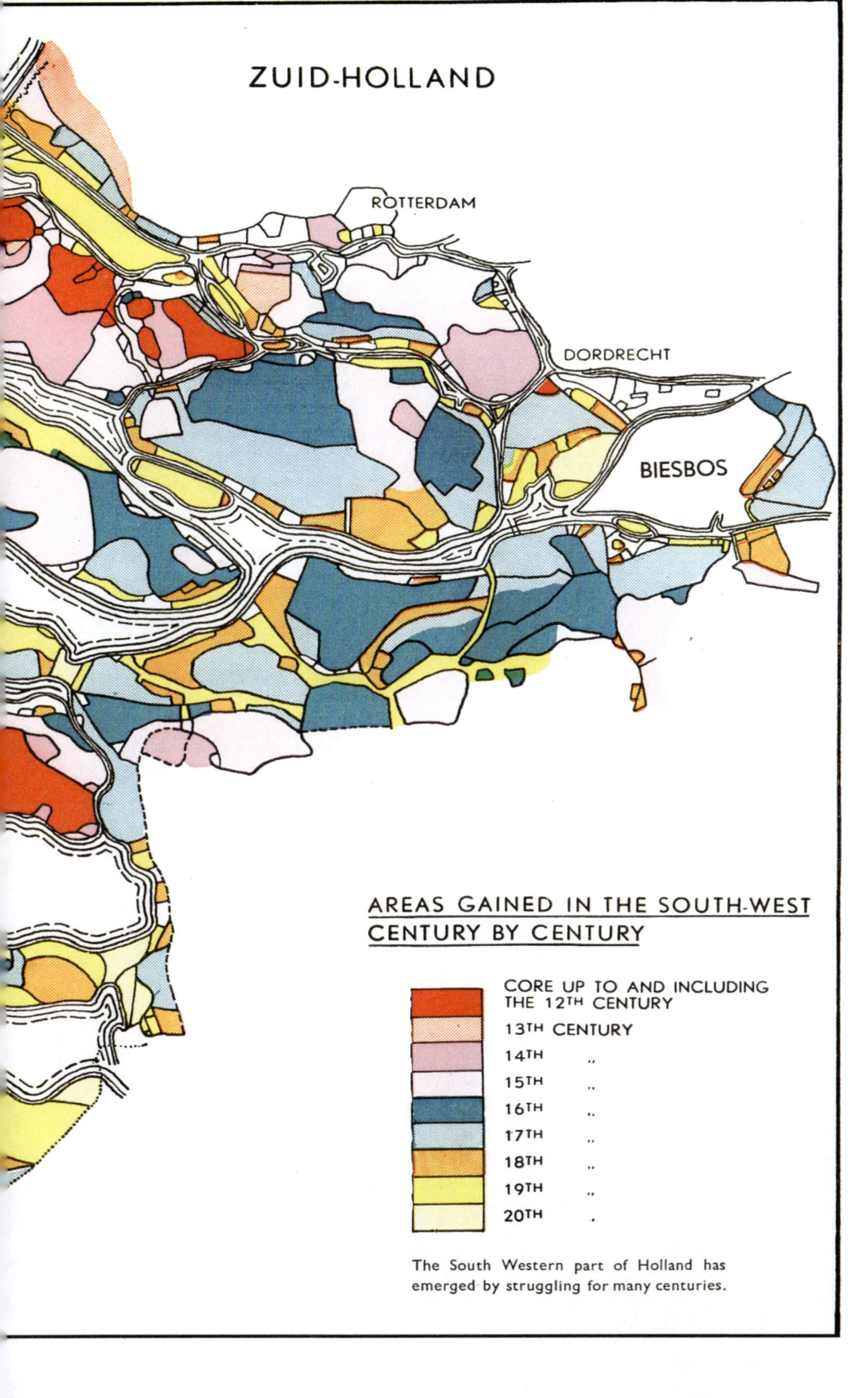

ZUID-HOLLAND

ROTTERDAM

DORDRECHT

BIESBOS

AREAS GAINED IN THE SOUTH-WEST CENTURY BY CENTURY

CORE UP TO AND INCLUDING THE 12TH CENTURY
13TH CENTURY
14TH   ,,
15TH   ,,
16TH   ,,
17TH   ,,
18TH   ,,
19TH   ,,
20TH   .

The South Western part of Holland has emerged by struggling for many centuries.

water of the sea, which is about 2 or 3% heavier than the river water itself. This difference in specific gravity causes a flood current near the bottom, so that when a ship of a fair draught is sailing into the sea, its speed falls. This is so because the 'devil' is holding the ship by its keel! Or on the other hand, the underflood may draw the ship into the river much more quickly than the captain expects.

This uncanny underflood, caused by the heavy salt water, also explains why the river sands cannot pass into the sea. Fine silt, suspended in the upper layers of the river may be able to reach the sea, but the heavier grains of sand which remain near the bottom, cannot. Instead, the underflood brings sea sand from the outside into the river, and it never comes out again except by dredging.

The people of the middle part of the Rhine had therefore a comparatively easy task, whereas the Dutch, at the lower end of the river, had a difficult one, and still have. While the Middle Rhine was being corrected, our difficulties increased considerably, because we got the surplus sand which the Germans drove down to Holland. Our German neighbours were able to boast that their part of the Rhine was in fine condition and everybody could see that the Dutch section was not. Hard thinking was needed once more. The frequent river flood disasters, .the constant wish to provide more depth for shipping and the obligation we had accepted at the Congress of Vienna, urged Waterstaat to frehs exertion.

What was the fundamental trouble? It was that our section of the Rhine could not be forced to bring its sands into the sea. Well then, let the rivers bring their sands into the sea-inlet known as Hollands Diep, which is 40 miles inland.

This idea was carried out. For the main branch of the Rhine, the Waal, a new outlet was made into the Hollands Diep, and for the river Maas a similar new mouth was also dredged to the Hollands Diep. This restored conditions more or less to what they were before 1213. The first new branch had a length of 11 miles, the second 22 miles. The widths were about 1150 to 1800 feet. Moreover, all other rivers in Holland were corrected.

Though we got some more depth by means of these improvements, the problem of the transportation of sand by the rivers themselves remained unsolved. Lowland rivers near their sea mouths as a rule cannot be helped in their task of carrying fluvial sands into the sea.

It was again the steam dredge which solved this 'insoluble river problem'. Without this dredge we should still be experiencing the same river troubles. If this, *the most important of all engineering machines*, had not been invented, Holland would still be a poor country. Our harbours would be third-rate, our rivers would be practically unnavigable and unmanageable and we should be a poor overpopulated nation with little trade and significance.

Up to the year 1885, when most of the Rhine had been corrected, the total

WHERE THE RHINE AND MAAS JOIN (Biesbosch)

Both rivers, shown here, are artificial. After having joined they are crossed by two bridges. The ancient course of the Maas, dammed in 1270, is indicated by a dotted line. In 1421 all the land shown was inundated and lost. Since then, silting up has produced much new land again, though several creeks remain. *(Photo K.L.M.)*

cost of that work had reached about £ 14,000,000, including the German part of the Rhine. This figure seems high, but it is little compared with the huge shipping which resulted. On the Dutch section of the Rhine the annual quantity of goods transported averages 40,000,000 metric tons, whereas on the Maas 10,500,000 tons were transported annually. The Rhine has become one of the greatest inland shipping routes of the world; its navigable depth in Holland is often 16 feet and nearly always more than 10 feet. The barges are generally 1000 to 2000 tons each, but there are some of 4300 tons. Coasters run up the river as far as Basel in Switzerland.

The difference is great when we compare the depths in this splendid river with those which King William I found in 1821, depths that would not allow his yacht three feet draught.

16. *Canals for Inland Shipping*

When some years after the Napoleonic wars the Railroad knocked at our door,

THE MEANDERING MAAS, which so often inundated large areas, has now been shortened and harnassed. The old loops are still there. The second river in the background is the Waal, the main branch of the Rhine. At one point they come close together. The Maas is now navigable for 2000-ton ships, the Waal for ships of more than 4000 tons. *(Photo K.L.M.)*

this new means of transportation was not considered practical in a country such as ours, because the land was intersected by thousands of canals and wide ditches. Moreover, it was felt that it would be impossible to compete with the system of efficient water transportation. Had not we the much envied 'Trekschuit' in which everybody could reach his destination comfortably and quite quickly too? The horse pulled the ship through the canals and the driver-boy blew lustily on his horn while the passengers smoked their clay pipes in rural peace. Who would prefer a dusty, smoky, clattering affair on rails?

Yet the Railroad came, though late. Many were the movable bridges which the engineers had to make across the canals. Of course, water traffic would retain its ascendancy over the railroad, the officials said. The bridges would be temporarily closed when a train was due and after the train had passed the bridges would be open for the water traffic again as soon as possible. Shipping would not need to suffer from this new fad.

This preference was not maintained. Ships find closed railway bridges now and they are opened for a short moment when the ships ask for it. But water transportation is still of great importance in Holland.

106

In the course of the next few years the Rhine west of Arnhem will be canalized. The first of the three barrages, located near Hagestein, is ready now.

Every village was supposed to lie on a canal. In the low-lying parts of the country it was even customary for a house to be built both on a street and on a canal. This could be done by digging a canal in the centre of the street. Amsterdam, Leyden and Delft provide good examples of this method. In this way we acquired the picturesque town canals which are the joy of painters. In those days it was held that the turf used for the stove in winter, as well as many other commodities, could best be delivered by ship right to the customer's house in the town. The floating shops stopped at the front door to offer their merchandise. Barges were used, and in some cases are still used, for sewage disposal.

Where else could we find all the water the Dutch women of the marshes needed to wash the walls of the house, the yard, the pavement and even the very street? 'Cleanliness is next to Godliness' was accepted literally in the low parts of Holland. It transcended importance the comfort of or the respect for the individual. A Swedish Ambassador recounts how he once paid a visit to one

107

Navigation on one of the branches of the Rhine near Rotterdam. The dike with the road upon it serves as a foundation of houses as well. In the foreground a repaired ancient breach.   *(Photo K.L.M.)*

of the powerful Burgomasters of Amsterdam while the maid was busy scrubbing the hall. The sturdy lass beckoned to the Ambassador to get on her back, so that his shoes should not dirty her hall. In this way she took him to the Burgomaster – a man more powerful than any Doge of Venice had ever been, but not more powerful than his maid in her capacity of Preserver of Cleanliness.

The Frenchman Harvard visited the marsh country and wrote: 'The stables and cow-sheds are real apartments where 'messieurs les chevaux et mesdames les vaches' are superbly accommodated.'

Many of the canals were formed from ditches. In the lowlands of the west practically every ditch was in use as a canal by which hay and cattle were transported by scows. Even nowadays funerals are sometimes conducted by water. In these parts the use of land routes is an innovation.

Water routes always have been of great importance in the history of the Low Countries. A main international trade route went of old from England over Holland to the Baltic and Sweden (Birka, Sigtuna) by way of the Eider and Schleswig. A second branched off near Utrecht and Dorestad, two other famous ancient centres – the Rhine.

108

Typical Dutch movable bridge across a canal, allowing quick passage of ships.

At this knot of great shipping routes the Dutch became the experts in transporting goods at an early date. In accordance with this, two periods can be observed in Dutch history in which their skippers obtained the hegemony of the trade of northwestern Europe. The first peak of their trade history was in the 7th and 8th century (centre Dorestad), the second in the 17th century (centre Amsterdam). The Norman raids (800–1000) ended the first period of expansion, prosperity and glory.

When the Norman raids ceased, the chaotic conditions of Western Europe did not allow much trade. How to pass the 60 'legal' tolls on the Rhine and the many hawklike robber barons perched in their nests above this fine river? The sea was full of robbers too. In order to avoid the pirates of the open sea the 'inner dunes route' came into existence, a series of rivers and canals by which one could sail from the Ems to Flanders (Bruges). The oldest canals were made in the north to provide a waterway between the Zuiderzee and the Ems. Part of it was called the Delf (Delfzijl). Another old canal connected Utrecht, built on the main mouth of the Rhine but silted up since about 900, with a new branch of that river. Some of these early canals have now also been silted up.

Slowly the shipping trade developed after the Norman raids. Money, used

Modern double locks for 2000-ton ships in the Amsterdam-Rhine Canal. Quick passing (5 minutes only).

extensively before Norman times, had been carried away to embellish the Norse beauties with bracelets and gold necklaces, tolls were the legal means by which all barons and kings had to live and with which they fought their wars. Rivers and canals could be easily 'shut off'. So land routes got the preference, they mostly led to the old centre, Paris, which was far away. But 'navigare necesse est' and Bruges, in the south of the Low Countries, soon became the great harbour (after 1100). The Normans had cost us three centuries and their leaving us alone was the beginning of new prosperity in trade, which did not leave the Low Countries until after 1750, and now led to such exceedingly great heights. Especially after the low 'Wapelinge' was being reclaimed, the inner dunes route and the many other canals which had been dug to drain the country, provided the means by which some big centre – Amsterdam – could spring forth. When Bruges' harbour became silted up, the centre of the Low Countries was ready once more to take over the leadership in shipping and trade. The southern (Flemish) part of the low countries remained powerful. Even in the 16th century Charles V could boast: 'Je mettrai Paris dans mon Gant' – I'll put Paris in my glove – Ghent being bigger than Paris.

Holland now possesses some 4800 miles of navigable canals and rivers. This is

110

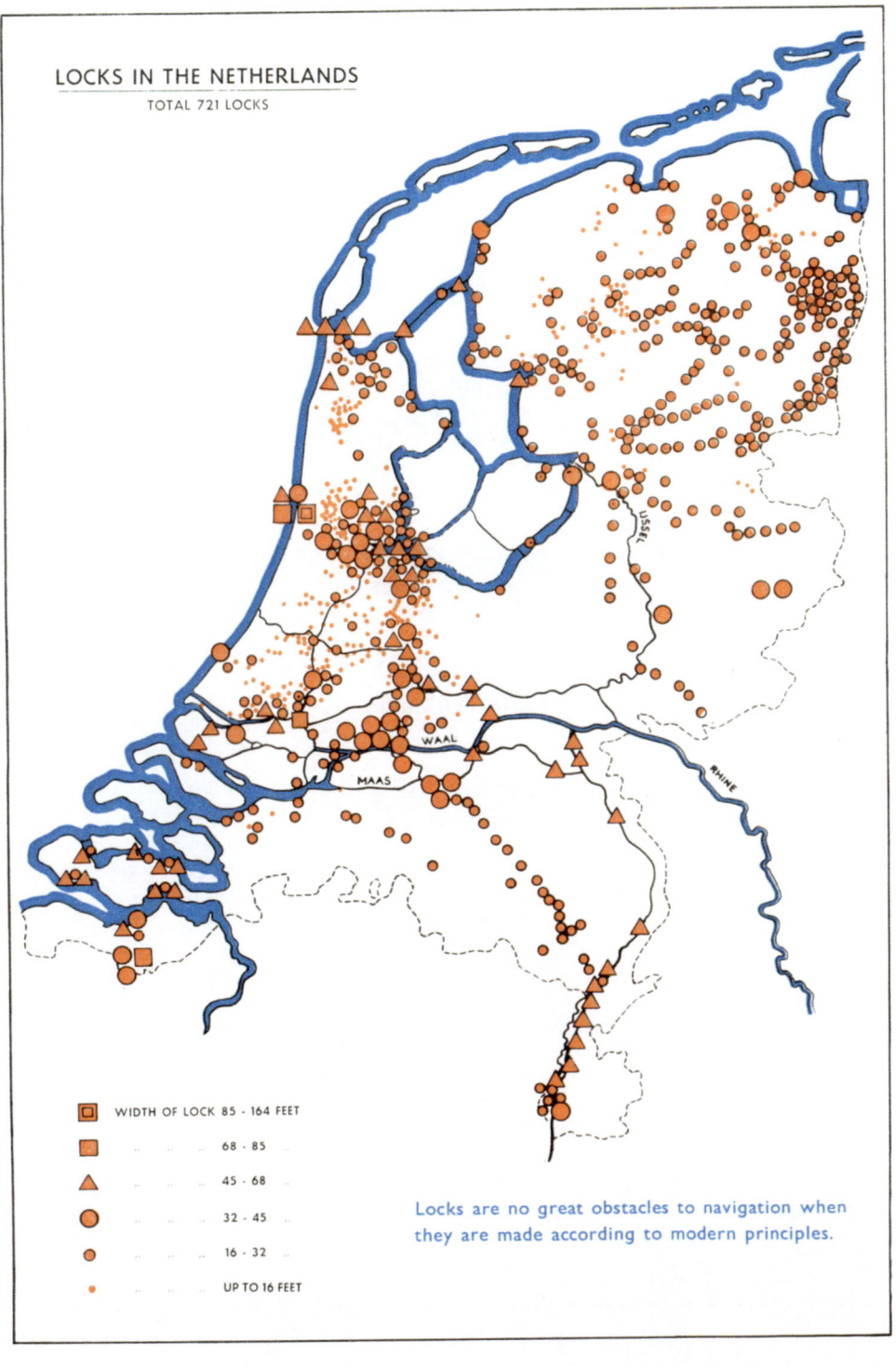

LOCKS IN THE NETHERLANDS
TOTAL 721 LOCKS

ISSEL

WAAL

MAAS

RHINE

WIDTH OF LOCK 85 - 164 FEET

" " " 68 - 85 "

" " " 45 - 68 "

" " " 32 - 45 "

" " " 16 - 32 "

" " " UP TO 16 FEET

Locks are no great obstacles to navigation when
they are made according to modern principles.

DIMENSIONS OF BIGGEST SHIPPING LOCKS
IN THE WORLD

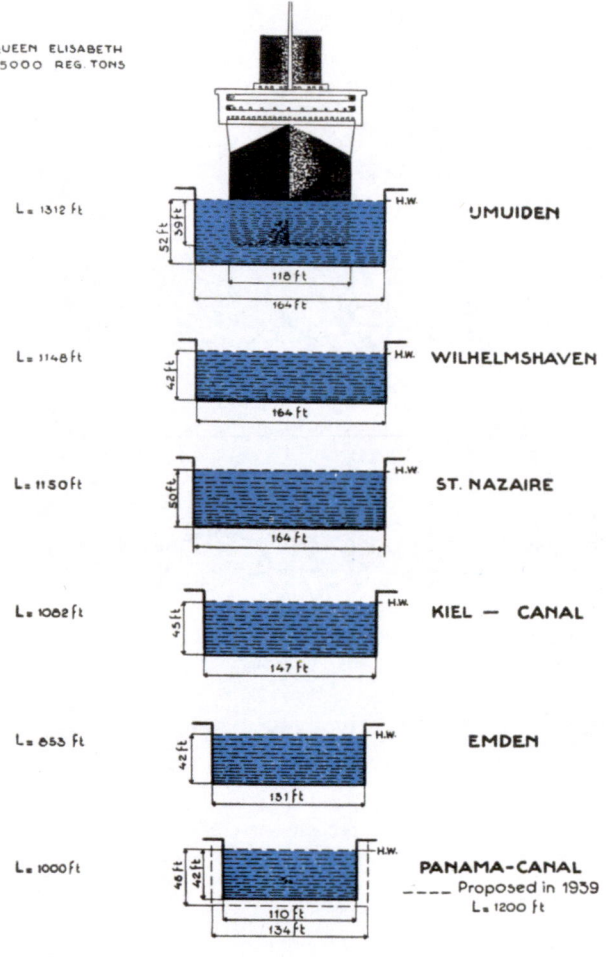

QUEEN ELISABETH
85000 REG. TONS

L = 1312 ft — IJMUIDEN

L = 1148 ft — WILHELMSHAVEN

L = 1150 ft — ST. NAZAIRE

L = 1082 ft — KIEL — CANAL

L = 853 ft — EMDEN

L = 1000 ft — PANAMA-CANAL
____ Proposed in 1939
L = 1200 ft

CROSS SECTIONS OF BIGGEST SHIPPING CANALS
IN THE WORLD

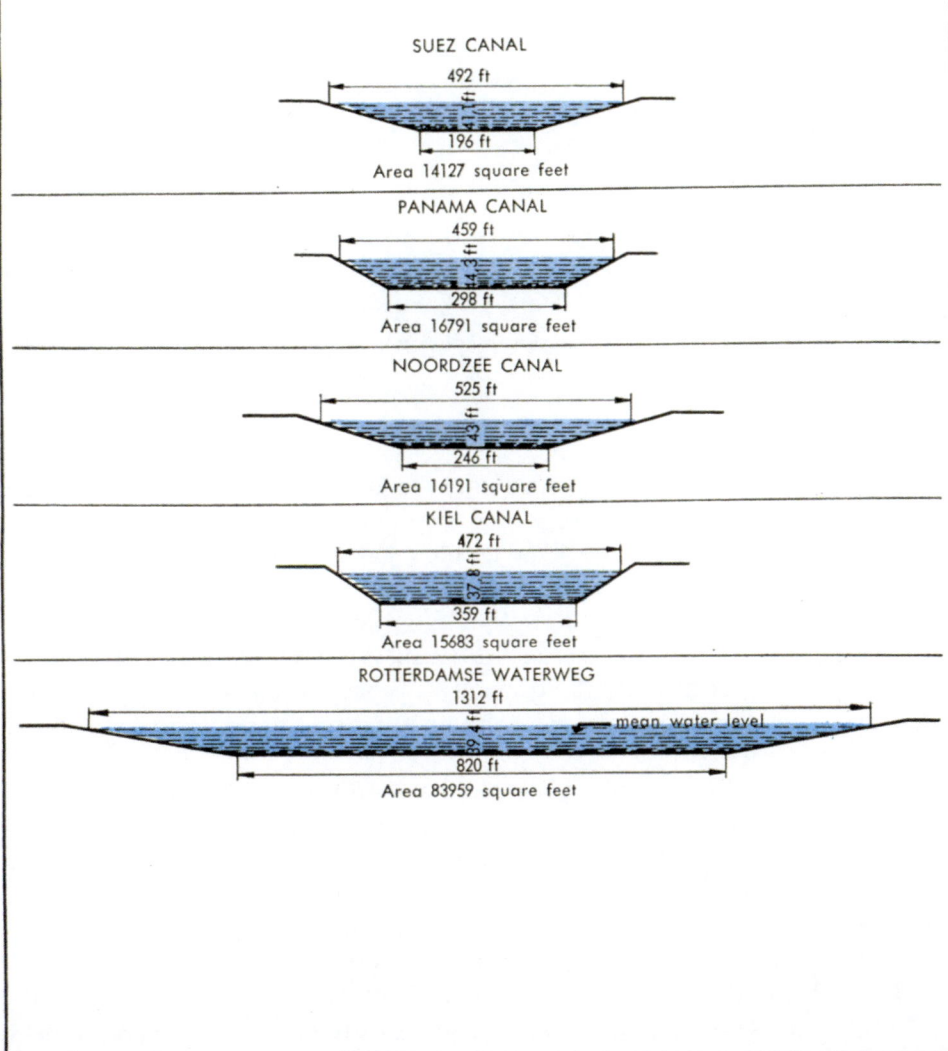

SUEZ CANAL

492 ft

196 ft

Area 14127 square feet

PANAMA CANAL

459 ft

298 ft

Area 16791 square feet

NOORDZEE CANAL

525 ft

246 ft

Area 16191 square feet

KIEL CANAL

472 ft

359 ft

Area 15683 square feet

ROTTERDAMSE WATERWEG

1312 ft

mean water level

820 ft

Area 83959 square feet

Weir and lock in the canalized Maas. The lock is for 2000-ton ships. *(Photo K.L.M.)*

about 48 times the average width of the country. Of these canals about 1500 miles are suitable for ships up to 400 tons and about 930 miles for ships up to 1000 tons. About 340 miles are navigable for vessels above 2500 tons, i.e. almost 3½ times the average width of the country.

When Henry Ford, the automobile king, saw all these canals his advice was to fill them in and to make roads of them. We have no intention of following his advice, for our fleet of inland vessels – about 20,000 of them – transport our goods, particularly mass products and heavy piece goods, in a cheap and effective manner.

In fact, one of the best ways to build a road in a low marshy land is first to dig a canal; some soft layers of peat are removed in this way before a body of sand is provided to act as a foundation for the hard road. This sand base often penetrates 50 feet into the soft layers, so that enormous quantities of sand are needed before the road can be built. It is for this reason that the roads in the low-lying parts of Holland are among the most expensive in the world.

In such soil the construction of canals is also extremely costly and difficult. There is a case on record of an attempt to dredge a canal which refused to increase in depth, no matter how much dredging was done. This lasted for days,

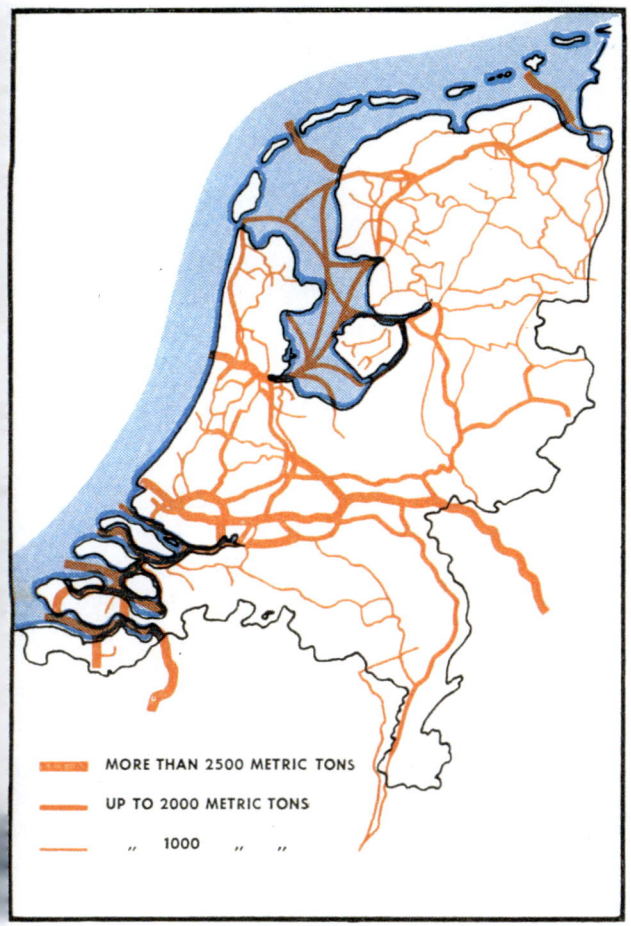

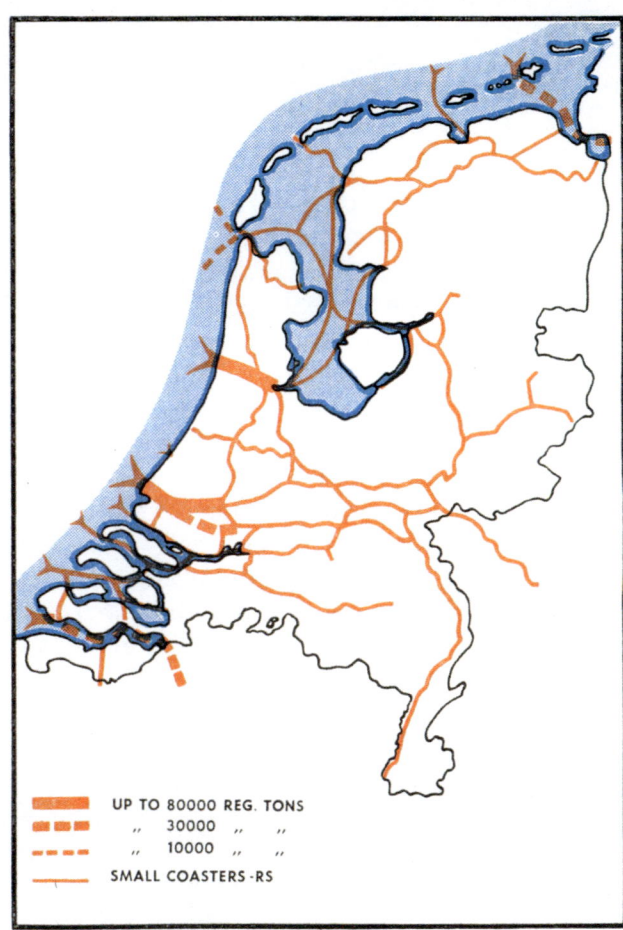

INLAND WATER TRAFFIC

The total length of the Dutch Waterways is 48 times the mean breadth of the Netherlands. Almost any village can be reached by water.

MARITIME TRAFFIC

The largest seagoing ships can come to Rotterdam and Amsterdam. Smaller craft can come to many other ports. Coasters can reach most towns of the Netherlands.

until suddenly a depth of 50 feet was found behind the dredger. The peat layer had had a tendency to float and had been completely dredged away.

The extent to which Holland depends on water transport is evident when we consider that in 1952 the transport of goods in Rotterdam alone was 43,409,000 metric tons. This means that working a full 24-hour day, 14 trains of 30 wagons each would be needed per hour. Water transport has no difficulty in coping with the Rotterdam trade, for a mere 75 barges of an average of 2000 tons burden would suffice per day. A single tug can easily tow 10,000 tons of goods, a performance equal to that of 20 to 25 heavy locomotives.

The development of inland navigation in the Netherlands is clearly reflected in the widening and deepening of the Amsterdam-Rhine canal since 1826.

In 1826: cross-sectional area 310 square feet, largest ship 500 metric tons
In 1890:   ,,   ,,   ,,   900   ,,   ,, , ,,   ,, 1350   ,,   ,,
In 1938:   ,,   ,,   ,, 2750   ,,   ,, , ,,   ,, 4300   ,,   ,,

115

The locks of this canal are now all 60 feet wide and at least 750 feet long. It generally takes no more than 15 minutes to pass a ship through.

A much-used waterway is that from the Rhine to Antwerp. It is largely composed of natural or corrected tidal waters, but a section of it consists of a canal, which also has large dimensions. Three adjacent locks at each end of the canal rapidly deal with shipping. The widest of these locks is 53 feet and about 20,000,000 tons are transported annually through them. One of the latest canals is that which has rendered accessible the industrial area of Twente in the eastern part of Holland.

It is beyond the scope of this book to discuss other canals of the Netherlands. Mention can be made only of the Maas canalization, which links our coal basin in the extreme southeast corner of the country with the central parts of Holland. The largest size of ship this canal can take is 2000 metric tons.

Of the 4800 miles of navigable waterways, 1000 miles are now suitable for coasters. The famous Dutch coastal trade has mainly developed in the northeastern parts of the country from the peat trade. The enterprising turf-skippers of those parts set out to sea with their small ships, reaching ports inaccessible to any other seagoing vessel. They became fearless seamen, who took their families with them to England, the Baltic, the Mediterranean, or even to America. Their ships were their homes. Successive generations of skippers were born at sea and if we examine the front leaves of their old family Bibles, we find that few of these people died a natural death. In 1858 the coasters from the northeast comprised 47% of all Dutch seagoing vessels and in 1954 it is about the same percentage. The tonnage, of course, was proportionately much smaller.

A fine book could be written about the history of these skippers. May it be done before it is too late. The women played an important role in this history. Listen to the story of a skipper as told to his pupils, when he had given up the sea in favour of the classroom: 'On the coast of Brittanny, a French vessel was on the rocks and the people squealed terribly. Father wanted to sail on, but the French were so desperate. So in that stormy evening my father, my brothers and myself set out in our lifeboat, while Mother had to handle the ship and the sails alone. It must have been hard for her to see all her menfolk leave the ship in that bad night. We were quite near the shore and there were foaming rocks, but Mother handled the ship so well that we managed to bring our Frenchmen on board in several trips.'

'Grandfather went to the Mediterranean with his ship, and had an old cannon on board against the Algerian pirates. When these pirates attacked, this blunderbuss was loaded with a handful of carpet tacks and gunpowder, and Grandmother heated her bowl of pitch to throw molten pitch into their faces with a big spoon.' When loaded the old ships had hardly any freeboard, but they were well covered with tarpaulins.

116

Not of least importance to our economy is the social life of the Canals, for it is to our stern, resourceful breed of skippers that we owe so much of our progress and prosperity.

It is remarkable that the Dutch skippers have been looked upon as ghost-skippers for such a long stretch of time. The Greeks said: 'They serve as ferrymen for the dead, and therefore they are the only people who are exempt from paying taxes.'

When we 'translate' Procopius' tale (6th century), we read that the Dutch skippers rise early, load their ships to sinking point, uttering a few outlandish words while loading, and are gone before dawn, sailing their too-heavily loaded ship very fast; supernatural forces must aid them in reaching their shores. It is well known that in Procopius' time the Dutch were handling the trade of the southern North Sea and the Rhine.

A late echo of this ghost-skipper story is the legend of the 'Flying Dutchman', the man and his crew who sail with full sails in a storm and enjoy it in a wild manner.

CHAPTER III

MASTERS OF THE FLOODS

17. *Reclamation of the Zuiderzee*

According to the experts now examining the soil which has been laid dry, about seven hundred years ago the Zuiderzee was a fresh-water lake. The salinity started to increase only about 1300 A.D. after the sea had gained wider ingress. This is in keeping with the sagas which speak of originally close contact between the Frisians on either side of the now broad waters and with a chronicler, who wrote that the water around the island of Marken was still fresh in 1240.

Shortly after its formation, the Zuiderzee became the scene of enormous developments in shipping, making it for a long time the centre of world trade, with fleets of up to 500 vessels coming and going. Later on, about 1800, it became a *Dream Sea*; for the last century and a half the name has called up images of baggy trousers and ringlets, and of quaint little towns, each with its own traditional style of dress, which have given Holland a unique reputation throughout the world.

This inland sea with its great history and capricious contours is now rapidly disappearing. In its stead one of the greatest gifts of Nature is springing forth. Out of the soil that once formed the bottom of the Zuiderzee sprouts wheat, – and is there anything better than wheat?

That question had already been asked by the legendary 'Widow' of Stavoren and she herself provided the answer. Stavoren, once the seat of the Kings of Friesland and perhaps one of the greatest ports of northwestern Europe during the 7th and 8th century, is nowadays an insignificant spot. It is called a 'grave-yard', but in the days when it was still a mighty mercantile town, the 'Widow', who was immeasurably wealthy, sent her fleet to all parts of the world then known. In her luxurious arrogance she ordered one of her skippers 'to fetch the best thing in the world at any price'. The simple skipper departed and after deep thought he bought wheat, the best he could find. Was not corn the life of man and did it not contain new life plentifully? Great was the anger of the proud woman when this cargo was brought to her, for she had expected silk, velvet and gold, or even more costly goods. 'You took it in over the starboard side, throw it out over the port side!' This wicked command had sad results. Since that day corn-like plants have grown near and in the harbour entrance and have gathered the mud. The 'Woman's Sand' was formed, which has since blocked the harbour. Stavoren passed into obscurity!

This tale is striking when we look at the rich fields of wheat which cover the

118

routes of the once-proud mercantile fleets of many ships sailing to Stavoren, Amsterdam, Kampen, Hoorn, Medemblik or Enkhuizen. The fate of Stavoren did not differ from that of the Zuiderzee as a whole. Instead of gold, velvet, silk, and other costly merchandise, it got wheat – rich and dense – standing on the bottom of the sea.

The history of the Zuiderzee, though short in years, is like a fairy tale. In its youth it was the Princess who had inherited the collective fame and glory of Venice, Spain and the Hanse. Later it became the 'Sleeping Beauty' who slept for more than a century. Today it has an entirely new appearance: a thriving cornfield, the most prolific on earth.

Whoever wants to see the strange but vanishing beauty of the Zuiderzee shores, should not wait too long. Our 'Sleeping Beauty' has been kissed awake by the Prince of Dredgers, and she is now rubbing her eyes in great wonder. – 'The unsurpassed and, indeed, often unmatched picturesqueness and historical interest of almost all the districts that bordered the old Zuiderzee are bound, in the nature of things, to suffer some change', writes the Rev. Edgar Brown and he adds: 'He who comes soonest to the lands along the Zuiderzee, now IJssel Lake, will be the more rewarded'.

Our quaint, dear, picturesque Zuiderzee, so full of ancient traditions – the real heart of the Netherlands – will soon become a healthy, super-modern farmer's country, in which everything shines with newness and fertility.

The reclamation of the Zuiderzee is one of the greatest works ever carried out by man for the promotion of human welfare. In its original design by De Lesseps, the appropriations for the Suez Canal were some frs. 200,000,000, or about £ 10,000,000. Up to the year 1885 the Rhine improvements cost about £ 14,000,000. The expenditure on the Rotterdam Waterway to date has been some £ 11,000,000 while the Panama Canal had cost about £ 55,000,000 up to the year 1915, the French works excluded.

By the time the Zuiderzee works are completed an expenditure of about £ 200,000,000 will have been incurred. This gives some idea of the magnitude of the task. The only comparable project is that commenced before the 1939–45 war in the Tennessee Valley. According to the figures published, the cost of the Tennessee project is of the same order of magnitude, but the Zuiderzee works were started 20 years earlier.

The object of the Zuiderzee works is mainly twofold:

1. To shorten the coastline of the lowlands with 200 miles by closing the Zuiderzee;
2. To create 550,000 acres of fertile land, and to provide a fresh-water reservoir in the heart of the country.

Obviously Holland wants more arable land, owing to the rapid increase of

Dr. C. LELY was the main promotor of reclaiming the Zuiderzee. When he was Minister of Waterstaat for the third time in his life, he at last saw his plan accepted.

her population. Now that machinery has become more powerful than formerly, Holland is continuing its age-old traditions in an enlarged fashion. To understand the necessity of a fresh-water basin, however, we need to know how insidiously the sea had commenced new tactics.

As a result of constantly-improving drainage, the land had in many places settled more and more below sea level, so that sea water penetrated into the polders underneath the dikes and dunes, poisoning the crops. At first we failed to notice this insidious approach of the sea from this unexpected quarter and consequently helped rather than opposed it.

It was particularly the horticulturists who sounded the alarm against this seepage. They proved that even in dilutions at which a brackish taste was hardly perceptible, the effect on fine horticultural crops was a diminishing yield. They gave as the maximum permissible limit 500 milligrammes of salt per litre, but our ditches already contained 1000, 3000 or even 10,000 milligrammes per litre. In periods of drought, when the plants needed a great deal of good water, the salinity was higher still.

Another cause of such seepage was that the farmers, ever inventive, had each erected a private gas factory, for the purpose of cooking and for lighting their farms. This gas was obtained from the earth (methane or marsh gas), and in its upward passage it carried the saline moisture of the deep layers with it. Also the

120

The old generation of Zuiderzee fishermen had
sad forebodings when their sea became a lake.
But more value of fish is caught now than when
the Zuiderzee was still an open tidal bay.

(Photo Maaskant)

farmers made deep wells in order to use the cold water from the subsoil for
cooling their milk in summertime. This cold water was brackish, too.

Especially in the brackish zones the malaria mosquitoes throve well and
brought fever – an additional reason for taking steps against the salt. It was
calculated that gas and cold-water wells alone caused the infiltration of many
hundreds of railroad car loads of salt daily into the low-lying part of Holland.
Measurements likewise revealed that every time it was used, the lock at IJmuiden permitted the ingress of not less than an amount equivalent to 100 or 150
railroad cars of pure salt into the country. As the locks are used many times a
day, much salt can be introduced in this way. Of course, a fresh-water reservoir
such as the Zuiderzee has become, enabling all this harmful salt water to be
flushed away, is of great value. Moreover, the gas and cold-water wells are
being choked now. Have we sufficient Rhine water at our disposal to keep our
country fresh? Now that the agrarian section of our people has become conscious of the danger of salt, the demand for Rhine water is constantly increasing,
so much so that in periods of drought we may anticipate a deficiency. We shall
therefore have to learn to be sparing in our use of Rhine water; too much is flowing unused into the North Sea. We shall likewise have to make it plain to the
French and Germans that they cannot indiscriminately drain the waste salt of
their potassium and coal mines into the Rhine. As a result of this the salinity of

121

ENCLOSURE DAM OF THE ZUIDERZEE

Fantastic hieroglyphics man has written in a sea. *(Photo K.L.M.)*

the Rhine has doubled itself in a few decades. We hope that this will be safe-guarded by International Law. The German officials themselves call the increase of salinity 'alarming' (erschrcckcnd), which gives us some hope for the future. There is a second method of combating salt from the sea, typified by the motto: 'shorten the coast, close the coast'. This is not a new principle, on the contrary it is traditional. In the year 1200 there was a long coastline to maintain, as many bays and creeks permitted the salt tides to enter the country. In 1840 we still had 1187 miles of salt coastline, but in 1930 the coast was reduced to 1040 miles. Now that the Zuiderzee has been sealed off we have no more than 860 miles. The coastline cut off by the Zuiderzee dike was 200 miles; the dike itself is 20 miles in length.

The great idea of turning this inland sea into land was mentioned first by *Hendrik Stevin* in 1667. He was the son of the famous mathematician Simon Stevin, who improved the construction of shipping locks; the son wrote a treatise: 'How to get rid of the Poison and Fury of the North Sea', but there are no technical details in it. The idea began to take a definite shape about 1840. Subsequently about a dozen schemes were published, until *Dr. C. Lely* ultimately suc-

122

ONE OF THE LAST GAPS IN THE ZUIDERZEE DAM

The closing of the last gaps had to be carried out as quickly as possible, hence the small profile of the dam just emerging above the water. After this struggle to effect enclosure, the dam could be made strong and high at leisure. Note the last tidal streams through the remaining gap.

(*Photo K.L.M.*)

ceeded in getting his scheme adopted in the midst of the 1914–1918 war by the Netherlands State Assembly. Lely, a tenacious, single-purposed man, was secretary of the Zuiderzee Association, a private institution which for many years had been directing public thought towards this end. It was the will of the people which eventually demanded the execution of the Zuiderzee reclamation. For despite the fact that this task imposed a burden on the present generation – future generations will reap more fruits than ours – the work has been enthusiastically supported by the entire population. It was not only the engineers and farmers who supported the plan, but the teachers in the primary and secondary schools have played an equally honourable part in making this dream a reality. It was a truly democratic enterprise, for the enthusiasm came from below, not from the top. Initiative, study and propaganda came from private resources.

Lely (1854–1929) was a graduate of the Delft University of Engineering. He weighed some 17 stones. His family was from North-Holland and was Baptist like Leeghwater's. Up to his 32nd year Lely was not very successful. In his career

123

The final enclosure of the Zuiderzee on 28th May 1932 was one of the great moments of the nation's history. *(Photo K.L.M.)*

at the 'Rijkswaterstaat' he never got beyond the position of the lowest imaginable grade of the lowest overseer category. He was unable to attain even the title of assistant engineer and could scarcely provide the means to keep his family in food and clothing. In 1886, at the age of 32, this highly qualified, able and good-looking engineer had been dismissed from the low job he had and had to lodge his family in his father's house, his earnings being only £ 60 in the whole year. 'But', said his father, 'I should not look for another job just now, particularly as long as the Zuiderzee Society has not been decided upon', for the Lelys were enthusiastic readers of any publication concerning the reclamation of the Zuiderzee. The library was blocked sometimes in this respect because the Lelys used the papers about the Zuiderzee plan so extensively.

What was the Zuiderzee plan in 1886 but a mere fata morgana. The Government kept itself completely aloof from it. Perhaps a few private persons would found a Zuiderzee Association and perhaps Lely might become a secretary in this castle in the air. But the Lelys, father and son, had a vision which transcended the Government's and they had some courage too. After starting a career with ten years of adversity many an engineer would have been happy with some small, permanent and regular job. Lely, on his father's advice, struck down

124

DRAINAGE SLUICES IN THE ZUIDERZEE DAM

One of the two sets of sluices in the Enclosing Dam by which the Rhine water is passed into the North Sea. 25 sluices and 2 large locks were constructed in the open sea before the dam itself was made. *(Photo K.L.M.)*

another path and associated himself with a highly uncertain cause the aim of which was to pursue the all but impossible. He became secretary of the Secretary of the Association.

By 1887 the Zuiderzee Association had collected £ 27,000 from private sources and it still needed £ 3,000, for which it applied to the Government. The Government replied in the negative and the Second Chamber of Parliament rejected the subsidy by 71 votes to 2. The Secretary now left this leaky boat, but the secretary Lely sat down behind his desk and drawing board and wrote and drafted his famous projects, which created such an impression on the Prime Minister Tak van Poortvliet that he made Lely a Minister at the early age of 36.

This was a great day for the man who just then was applying for an emigration ticket to Brazil, because he had seemingly no future as an engineer in his home country. On this occasion the father wrote him a letter in which he advised his son to continue humbly in the old Dutch style: 'Doe wel en zie niet om' (Do what is right and don't look back) 'and let come what may'. – On this day Lely was walking in the street and met a colleague, who inquired kindly about

125

ENCLOSURE DAM OF THE ZUIDERZEE

Height about 20 to 22 feet above mean sea level. The North Sea is on the left; the Zuiderzee–now the IJssel Lake–on the right.
There is room for a concrete road, a cycling track and a double-track railroad on the dike.

his leaving for Brazil. Lely answered as casually and humbly as he could: 'I
have got another job here, I have become the Minister of Waterstaat'. Lely
became a Minister three times but had to wait till his 63rd year before the Zuiderzee project was approved by the Houses of Parliament (1918). In the meantime he had acted – inter alia – as Governor of the Colony of Surinam.

Democratic Governments are, or were, great for developing tactical qualities.
Lely did not dwell unduly upon the Zuiderzee project during his first Minister-
ship; the scheme was considered by all experts as impossible. He merely appointed a committee for it and accepted the report. The second time he submitted a bill, but showed his faith by not pushing for acceptance, feeling that
the time was not yet ripe. But during his third period of office in the winter of
1916 a tidal wave swept along the borders of the Zuiderzee with disastrous
effect. And so it was finally the sea itself which tragically supported Lely to
realize his lifework. In this and in the wartime food shortage the country found
its incentives. In 1918 the Zuiderzee bill received a favourable vote from the
Houses of Parliament. Nevertheless Lely did not feel safe until 1925, not until

126

THE ZUIDERZEE DAM (Infra-red photograph)

Its length is 20 miles. The Wadden Sea is on the left, the new fresh water lake on the right. In the foreground is the first reclaimed polder, called Wieringermeer. In the background on the right a small part of the province of Friesland. Locks, sluices and harbour dams near the former island of Wieringen can be seen at the left centre. The tidal range north of the dam increased much as a result of the construction of the dam.   *(Photo K.L.M.)*

the Government had actually bound itself by some small first contract. He then knew that his ship was in a safe harbour and exclaimed: 'Now the dike will be made, because the Government cannot go back any more.'

Even then the experts hardly dared to believe that they would be able to shut off the tidal Zuiderzee, because the closing gap would have tremendous tidal currents, and a work of these dimensions had never been attempted before. Also the boulder clay which played such an important part in making the closing dike had not yet been discovered.

The figure of Lely is not interesting from the mere technical point of view. He never was a supervising engineer and he himself carried out no project. If he had been asked what was the place of technique in the development of the country he might have answered: 'La technique est aisée, la diplomatie est difficile.' At an early age, after the Cinderella years, he rose above technique and became not 'a master of the floods', but the master of the floods'. He was one of the comparatively rare types of engineers who, like De Lesseps, possessed

127

Willow mattresses being sunk with stones to prevent scouring of the new dike. In the background a floating pumping station, which pumps sand to make the dike.

(Photo Maaskant)

initiative and a broad view about the possibilities of technique as a servant for higher goals, technique not being a high goal itself. Lely was not a 'carpenter', he was an 'architect'. The officials of his Ministry of Waterstaat used to say: 'Lely's four years here give the Ministry enough to keep it going for another twenty-five years.' He prepared the way for the many keen technicians, who were now invited to tackle the splendid task which may take them more than half a century.

The construction of the 20-mile Zuiderzee dam in tidal waters required an unprecedented effort, greater than any imposed by previous feats of hydraulic engineering. The entire country shared in the great task and the joy was immense when finally, on 28th May 1932, the last gap was closed. Shortly before the final closure the sea had worn a depth of no less than 100 feet in the gap.

Previous calculations, extremely complicated in their structure, had disclosed that the tides outside the Zuiderzee dam would increase in strength. Professor *Dr. H. A. Lorentz* of Leyden University was the first to present tidal movements in the form of soluble formulae, and his forecasts were found to tally to within

ZUIDERZEE RECLAMATION

When the first reclaimed sea bottom fell dry, a dismal gray surface revealed itself in which the dredges had already made some rectilinear canals, leading to the pumping station. In the foreground the old castle of famous King Radboud, 'the Unpeaceful'.

(Photo K.L.M.)

an inch of the new tidal rise. As a result of these calculations, it was possible to predict and prepare for every contingency. Before the Zuiderzee dam was finished, the sea walls north of the new dam were raised in accordance with these same purely scientific principles. Mathematicians had now joined the farmers and the engineers in their fight against the sea.

Great use was likewise made of the inspiration of one of the Zuiderzee fishermen to use the heavy boulder clay, abundantly available at the bottom of the Zuiderzee. This tough diluvial clay has played a predominant role in the fight against the currents in the gaps and has helped greatly to keep expenses within reasonable limits. If we had not had this boulder clay, the difficulties would have been much greater. Admittedly, we often lost 50% or more of the sand and clay we had already deposited, but the rest remained and allowed progress to be made. By the time the last gap was closed the rate of progress of the work had assumed large proportions. In two months' time a gap of 3350 feet was closed.

A tragic feature of the reclamation of the Zuiderzee was the fate of those brave fishermen, familiar by their quaint dress. Was it to be wondered at that

129

DRAINAGE TUBES FOR THE ZUIDERZEE RECLAMATION

The heap shown here represents only a very small fraction of the drainage tubes needed. *(Photo A.N.P.)*

their hearts were filled with fear when the decision to pump their sea dry was made? Where could they earn their living? Were men who for a thousand years, or more, had been fishermen, now to become farmers almost overnight? The older generation of fishermen in particular could be seen sitting ashore, gazing with sad, ageing eyes at the sea which would soon cease to be a sea. Their world was to be annihilated.

Some of the fishermen became lock-keepers in the service of 'Waterstaat'; others learned a trade and many went to fish in the North Sea. Most of them, however, went on fishing in the sea, which had now become a fresh-water lake, for eel and other fresh-water fish, instead of for herrings.

It is interesting to relate how badly the fishermen fared initially and how their fears passed. At the outset it was just as if Nature wanted to bear out their gloomy forebodings; for hardly any fish could be caught any more. Instead of fish Nature sent a plague of mosquitoes, a plague so terrible that every house, every tree, every field and every human being who ventured out of doors, became covered with a grey layer of sluggish mosquitoes, like a covering of dirty snow. These pests did not bite; they rendered it impossible to look out of

130

the windows or to drive a car, for they obscured the view. If the windscreen wipers were set on motion, they only crushed the bodies of the mosquitoes, spreading a thick blubber over the windscreen, which no eye could penetrate. In some parts mischievous boys collected handfuls of mosquitoes, like snowballs, from the streets and window sills and threw them at passers-by. Many of the old fishermen saw in this plague the proof of their belief that God does not permit interference with his plans. Egypt, too, had had its plagues.

We had violated the ways of Nature, but again Science provided an answer. The fishery experts put forward a solution, remarkable for its simplicity. Their advice was to use the locks in the Zuiderzee dike at nighttime, preferably about midnight. The elvers, having come all the way from the depths off the coast of Florida, and being avid for freshwater food, now passed the locks of the sealed-off Zuiderzee by millions, there to devour the mosquito larvae, for the eel is a nocturnal hunter and sleeps in the daytime. They scented the food they desired, as it were, and at night had waited in squirming multitudes to gain admittance through the locks. But in the daytime they all lay in the mud with their heads just sticking out, asleep and unaware when the way to their quarry was open.

This simple use of natural means saved millions of guilders. The fishermen got their eels, fattened on mosquito larvae. The Egyptian plague which had assumed such frightening proportions, disappeared. The biological equilibrium, disturbed so roughly, readjusted itself.

The construction of the Zuiderzee dam was a magnificent achievement. Its crown is 20 to 22 feet above mean sea level and the width of the base at the bottom of the sea is about 600 feet. Driving to Friesland over the concrete road built on its surface, one gets a sense of the real magnitude of the undertaking.

Yet, the completion of the dike was only a small part of the task of reclaiming the Zuiderzee. The sea had become a lake, but in this lake some hundreds of miles of secondary dikes had to be constructed for the polders. After this the four polders, totalling an acreage of 550,000, had to be pumped dry. A new lake, called IJssel Lake, which will have an area of 270,000 acres, remains to provide fresh water for the low areas of Holland. The branch of the Rhine called the Gelderse IJssel, which flows into this new lake, keeps it fresh and the superfluous water from this river is released by 25 sluices having a total width of 1600 feet. These sluices were constructed in two groups in two pits in the open sea before the actual construction of the Zuiderzee dam. During the period of building, the sluices were surrounded by a dike to keep out the storm floods. Two shipping locks, both capable of passing 2000 ton ships, were likewise constructed in the open sea, surrounded by circular dikes, which were dredged away later on.

The first of the four Polders, covering some 50,000 acres, or about 9% of the whole reclamation scheme, was surrounded by secondary dikes and pumped dry even before the completion of the main Zuiderzee dam. To ensure the safety of the low-lying sea bottom, the secondary dikes are also of robust construction. If by some mischance the main Zuiderzee dam should be pierced, these secondary dikes have been designed to be capable of withstanding the North Sea tide. They stand about 18 feet above mean sea level.

The first polder in the former Zuiderzee was by far the largest in the world, yet still the smallest of the four to be constructed. The second, with an area of 120,000 acres was pumped dry in 1942. A start was made with the third in 1950; in the summer of 1957 an immense mudplate of 133,000 acres dried off. The complement of this polder – about 100,000 acres – is under construction now. An idea of the size of the first and smallest of these polders can be formed by a few figures. In it 150 miles of road were constructed, as well as 12 miles of canal 9 feet deep and 26 feet wide; 59 bridges; and 2 pumping stations with capacities of 120,000 and 160,000 gallons per minute respectively. The total project of the Zuiderzee will probably be completed between 1980 and '90, by which time the work will have lasted more than half a century.

The works attract many thousands of visitors annually, particularly farmers, who want to see the land with their own eyes. For every new farm leased out, there are three to four hundred applicants. The State administers the property and builds the modern farmhouses, schools, churches and even whole villages. It was pleasant to see how in a few years the originally bare and inhospitable land had changed. The large farms, which had first stood naked and alone in the landscape, were surrounded now by shrubs and flowers. Trees had been planted along the roads and here and there even a wood had sprung up.

The rich, waving corn was the polder's splendid summer raiment and the tractors and clanking mowing machines were witnesses of healthy, joyful work. The trilling of the lark proclaimed the joy of Nature. – This lasted only until 1945, then the whole glorious polder was wantonly inundated by the Germans. All buildings were greatly damaged and all plant life was destroyed. Yet, in 1946 new crops again were reaped.

Many foreign visitors who come to Holland are struck by the fact that the bottom of the sea on which men walk, live and work, so rapidly loses the impression of having been captured from the sea. It soon assumes the same character as the rest of the low-lying parts of Holland. This is natural, as it lies at the same level. A Czecho-Slovakian author writes: 'A typical polder can be recognized by its exceedingly tidy, lush and pleasing appearance and by the fact that there is nothing about it which recalls so stirring an event as the struggle between man and water'.

This is no mere witticism, but the truth. To the ordinary visitor, the polders are distinctive by the great care with which the entire landscape has been laid

132

out. It is not the low level of the land which attracts attention, for this is everywhere below sea level.

It is only natural that the population is pleased with the new soil presented to them by the engineers. What a fruitful soil it is! – But what is the view of the foreign hydraulic experts who realize what such great projects entail?

The President of the British Institute of Civil Engineers, *Sir George Humphreys*, points out that the greatest possible use has been made of the materials available in situ: boulder clay, sand and brushwood. Such clever and economical work deserves, he says, special mention. We appreciate this praise. Vierlingh would hardly have wished to accept us as his descendants, if we had ignored his rule to work 'at little cost'. He said: 'It is not enough to make dikes and dams, but they must be made cheaply for the general profit'. And again: 'By his Grace, God has given us these materials to make dams: brushwood, turfs, stones, clay, straw, reeds, etc. and you need not seek others, but use the materials you have artfully.'

In Egypt we worked with papyrus, in England with blackthorn and in China with millet stalks. The Chinese also planted out willow and today it is a commodity sold universally in China for hydraulic works.

As for the Chinese, we have great admiration for their ancient hydraulic methods, for the way their dikes are constructed and for the manner in which they close gaps, though their methods differ from ours.

18. *Walcheren*

Sometimes when one has at last succeeded in completing a particularly onerous task, one finds oneself suddenly confronted with an even tougher job. The Zuiderzee had been sealed off only 12 years, when the problem of closing the breaches in the dikes of Walcheren confronted us. The scope of this work was not so great as that of reclaiming the Zuiderzee, but technically it was more difficult.

In October 1944, the Allied Supreme Command deemed it essential to inundate Walcheren, which at the gateway to the Scheldt, had been heavily defended by the Germans. The plan was completely successful. At four points the Allied Air Force bombed large breaches in the dikes and the sea poured in. An area of 40,000 acres was flooded with salt water and the population fled, some to other districts, others to their attics.

Soon the breaches made by the bombs assumed formidable proportions, as the tides ran in and out. The flooding became worse and worse. What was left of Walcheren was now conquered by the Allies, while the war was still in progress, there could be no question of closing the gaps.

Would it ever be possible to make Walcheren habitable again? So often in our history we had had to surrender ground even more extensive than this island. Often the sea had remained inside, and that might be the fate of Walcheren, owing to the great tidal rise at its shores.

As soon as Holland had been liberated south of the Great Rivers (1944) 'Waterstaat' engineers in the liberated zone were sent to Walcheren. They were unable to reach the breaches, for the area was full of landmines and there was not a single boat to be found. It was only when someone eventually discovered a rowboat somewhere on the mainland, that it became possible to inspect the breaches. The sea had not been idle. Every day the channels had penetrated further inland and the breaches had become wider and wider. If the island was to be saved, where were we to find the workmen and tools to do the job? The population had left the island or was living in appalling circumstances. It would therefore be necessary to bring labour from afar, but where could they be housed? Similarly, there was no dredging material left in the liberated part of the country.

A start was made, though progress was slow. The first dredge was raised from the bottom of the sea inlet, where it had lain submerged. The Allies gave every assistance within their power. Switzerland sent wooden sheds; Belgium granted a credit for the purchase of stones; dredging equipment was sent from Great Britain and Belgium.

It was not possible to undertake the task seriously, however, till the North of Holland had been liberated. This took place on 5th May 1945, and once the mines had been cleared, the powerful dredging fleet belonging to the Association of Contractors who were carrying out the Zuiderzee project, the M.U.Z., sailed out to the southwest to save the island of Walcheren. This was in June 1945. Soon there were no fewer than 312 units of floating equipment at work: – 14 dredges and suction dredges, 135 barges, 71 tugboats, 73 landing craft and freighters, 19 large and small floating cranes. There were 52 draglines and bulldozers in operation. Also 73 barracks were built to accommodate the labourers. The breaches were now almost of a year's standing, so it was more than time to put a stop to the process of erosion. The total width of the four breaches was now 3000 yards.

Those who viewed them in the late autumn of 1945 considered the task of closing these gaps before the winter extremely difficult. The great tidal difference between high and low water, about three times as high as had been encountered in the Zuiderzee project, gave rise to forces theoretically 9 times more powerful. The currents in the breaches had already attained very high speeds and winter was about to come. We had already found it a difficult task to close the final gap in the Zuiderzee dike and even with such a small tidal amplitude a hole 100 ft. deep had been scoured. But now we had four much wilder gaps to close – not in summer, but in wintertime.

134

1945 on the flooded isle of Walcheren.

What can one do when confronted with such a task? Only this – set to it with a will, work, improvise and apply initiative. One learns from mistakes and sticks to the task, hoping for the best. Posters throughout the country bore the motto *Walcheren must be laid dry!* But to be truthful, in the beginning it was far from clear to the contractors and engineers how this was to be accomplished. They attempted the tried method of brushwood construction, which had stood them in such good stead for many centuries, but it was evident that this method could not effect a complete closure of the breaches, owing to the powerful currents. Just previous to this a particularly heavy clay had been found some 60 miles from Walcheren. This clay was dredged and transported to the breaches by means of 20 tugs, which maintained a steady supply. But even this was not enough to close the gaps.

Eventually, in desperation, after previously providing brushwood mattresses to prevent scouring, so-called 'beetles' (landing pontoons) were scuttled in the breaches and when this was found to produce good results, not only these concrete 'beetles' were used to close the gaps, but large ships, Mulberry pontoons and much other war material placed at our disposal by the British Government. In all, 70 concrete landing craft and other vessels were sunk.

The largest of these were 3000 tons and had a length of 200 feet. Good use was also made of tangled masses of torpedo netting, which were thrown into the breaches. The labourers gave the last ounce of their strength and one of

135

Enclosure of one of the gaps of Walcheren by means of scuttled ships. The fierce currents have broken the ship and find their way underneath it.

the chief wits gave an interpretation of their sentiments when he said: 'If a herd of white elephants floated here tomorrow, I should not be in the least surprised. I should only ask, where must I fasten the anchor rope on them and in which breach shall we sink them'.

At last the first of the gaps was closed, but the next springtide the new closure dike which had been so dearly bought was destroyed once more. After a few of these tides the hole was as deep as before. Again, success seemed impossible; the currents insuperable. Yet at long last Walcheren *was* reclaimed, after our modern technical resources had been strained to their uttermost.

In other countries, too, strong currents have been dammed, but mostly there has been a rocky bottom which could not be worn away. We had to work with a loose sandy bottom, which started to shift even at a speed of only half a knot. The difficulties we encountered with Walcheren have only increased our admiration for our ancestors. It seems incredible how with their primitive means they succeeded in closing gaps which must have had tides as fierce as ours rushing through them.

Our success in closing the gaps of Walcheren is partly due to the assistance rendered by our Allies. Also the experience gained in the Zuiderzee works was a factor. Calmly and with great navigational skill the heavy ships laden with

Strong tides in the breaches of Walcheren, causing waterfalls of more than 4 feet. The soil consists of loose sand, upon which willow mattresses have been laid. *(Photo Henning)*

stones or sand were guided to the appointed spot during the few minutes the tide took to turn and were then scuttled by means of dynamite. The impetuous and wild element was the water, which kept roaring through the breaches. Man himself was calm and 'made use of the time' as prescribed by Vierlingh, at the right moment. It was impossible to advance during the spring tides, progress could only be made during the neap tides.

An important factor was that we did not postpone action longer than necessary. The system of channels which the floods had started to develop, had only been formed to a relatively small extent. The destruction of the soil of the island was but a small percentage of what it would have been if we had allowed the process of erosion in these channels to continue. As it was, about 16,000,000 cubic yards of water were being washed through the breaches at every normal tide. It could have become twenty times this figure if allowed to develop.

In this trial of strength, not much attention could be paid to economy. When after the greatest of troubles the fourth and last breach – more difficult than any of the others – had been closed and the new dike shortly afterwards collapsed again, so that soon we were sounding depths of 80 feet in the breach, no fewer

137

than thirty thousand tons of stones were poured into the gap. Moreover it took 39 scuttled ships to fill this gap. This reminds us of the brief letter from Churchill in which he ordered the construction of an emergency harbour on the French coast, saying: 'and no arguments about it'.

What was done is a proof of our deep appreciation of the work of our forefathers. Whatever has been captured from the sea must not be surrendered to the sea at any price. Was the whole island worth the money spent on its reclamation? The expenditure involved was about four times the value of the cultivable land on the island, but the State cannot work on such narrow conceptions. The social and even the general economic value far transcends the restricted private value.

Walcheren emerged from the strife torn and devastated. The salt water killed all forms of plant life, millions of mussels grew on the dead shrubs and trees. The dead wood has now been cleared away and the first young trees have been planted. Since Walcheren was laid dry, the farmers have returned like swallows to their nests after winter, and ditches which had silted up with sand have been re-dug. The houses have been rebuilt and the badly-damaged roads repaired. True to the age-old motto of Zeeland, 'Luctor et Emergo', Walcheren has re-emerged above the water. After a few decades, when the glitter of newness has passed off, the sufferings endured and the struggle for its recapture will be but a memory, to be recorded in our historical archives, with the other catastrophic episodes of the past.

19. *Scientific Investigation*

When in the year 1213 Count William I and his men sliced some 30 miles from the lower mouth of the Maas by damming the river at two places and surrounding the low-lying land on either side of the cut-off river with dikes, they accomplished a huge task for those times. It was perhaps not a work which was scientifically justifiable.

We have great admiration for the daring of those men and for the fact that as early as 1213 they managed to seal off a river 400 yards wide, despite its powerful tidal currents. It must be admitted, however, that the resultant catastrophes were of alarming proportions. Nowadays we cannot but regard the creation of the 124,000 acre polder of 1213 as a rather reckless deed. For 207 years the struggle against Nature, which had been so ambitiously challenged, was kept up, but eventually this polder with its 65 villages and 10,000 inhabitants was submerged in a single night.

In their venturesomeness they had underestimated the forces of Nature. Technically they were already capable of subjugating the rivers to their will, but in 1213 scientific knowledge was still too undeveloped to gauge the effects of their deeds. Vierlingh's saying that 'whoever uses force on water shall have force exerted on him by water' is a lesson which we have learnt slowly.

Knowledge is power and ignorance is weakness. It has taken us a thousand years to acquire the knowledge which in our times enables us to live more safely than in bygone days. When we review the lessons taught us by the sea and rivers in the course of the ages, we can claim to have now learned the following principles:

1. HIGHER DIKES CAUSE HIGHER FLOODS

In low, non-diked country, the floodwaters spread without reaching a high level. When the first dikes were made, it was therefore thought that they could be kept fairly low. Later it was noticed that the higher the dikes, the higher the floods. Nowadays we know that so long as the dikes have not reached their critical level, which can be calculated, the heights of their crowns will determine the highest flood levels. For when a flood overflows a dike it stops rising, as the flood waters are utilized to fill the polder.

Since 1920 the 'critical level' has been calculated very accurately by using exact mathematical tidal formulas.

2. SALT BURNING

In the early days large incomes were earned by digging the salt-laden peat and burning it until the salt was left. Salt was used in great quantities for exporting herring and other fish. This burning of salt resulted in a destruction of the land by pitting it with holes, which became lakes. Once there was a breach in the sea walls, the sea destroyed these pitted areas easily, so that the lakes became sea inlets. Such lands were quickly beyond repair. This 'selbernen' or salt burning was forbidden at an early date, but went on clandestinely for a long time.

3. SUBSIDENCE BY SHRINKING

With better methods of draining, several diked-in soils began to settle some yards. If in such a settled land the sea broke through, the water stayed in the polder, and tidal movements at once became powerful. Under such circumstances a breach in the dikes is very dangerous.

4. GEOLOGICAL DEPRESSION (ELEVATION OF SEA LEVEL)

This slowly-approaching danger is difficult to gauge even in these days. Presumably there is a certain fluctuation with rapid and slow risings of the sea level. In the long run it is a danger against which we have so far no preventative. In its present evolution its magnitude is of the order of some inches per century. Geologists maintain that the central part of our country has subsided already about 8000 yards! The melting of the ice at the poles by about 25% would raise the sea level to such a height that only a small and infertile part of the Netherlands would remain. A 10% melting would be disastrous for the lower half of the country.

5. THE ROT OF FOUNDATION PILES BY BETTER DRAINAGE

All towns in the districts with a watery soil such as Amsterdam, Rotterdam, Gouda, Leyden, etc. are built on long wooden piles. The Amsterdam town hall, built in 1660, has 13,659 piles. When a lake is drained in the neighourhood, or also when the southern part of the Zuiderzee is going to be drained, the water table under Amsterdam is lowered and the tops of the piles rot. The whole town would collapse if the ground water level was not kept at its original level by artificial means.

6. THE PILEWORM PLAGUE

This threat to the country when the pile worm sponged the heavy wooden seashore defences, a threat impossible to foresee, was scientifically tackled by two simple men, who were the forerunners of the principle of 'fine, coarse and very coarse', i.e., that erosion by the action of waves on sand and clay is prevented by a cover of gravel or rubble and that the latter in turn can be safeguarded by the provision of large and heavy stones.

7. RISE IN RIVER LEVELS

The building of dikes along the branches of the Rhine and the Maas caused the bottoms of these rivers to rise and has resulted in a series of major disasters during many centuries. Dredging brought the only possible solution.

8. THE UNDERFLOOD

Owing to differences in specific weight of salt and fresh water there is a residual bottom current from the sea landwards. This current tends to cause a choking of the river mouths: it opposes the release of river sand into the sea and it causes a bar at entrance. Dredging provides the only possible solution.

9. SALT INFILTRATION

This subterranean activity on the part of the sea has naturally been in progress for a long time, but it has only been discovered recently how great the damage to agriculture and horticulture is.

10. SALTING OF THE RHINE WATER BY THE FRENCH AND GERMANS

If this salting is not checked, the low-lying parts of Holland will not maintain their present productivity, for the salt acts as a dangerous poison.

140

11. LOCKS ARE NO OBSTACLES TO SALT INFILTRATION

This discovery is one of recent times. Very much salt is admitted into our country by means of locks.

12. SILTING-UP OF HARBOURS

This had been one of the greatest menaces of the sea, but a means has been found to combat this evil: the dredge.

The troubles outlined above seem partly open to easy detection and cure, but only in retrospect! Our ignorance would appear to have been profound and our road to wisdom beset with difficulties.

The ways of Nature can be learnt only by bitter experience, or by scientific investigation. The former is a costly, difficult and tedious process. We have long trodden the hard road of bitter experience and our eyes are now turned to the shorter and more promising path. If only we had known all we know now! We should not have had to surrender a large part of our land and some of our projects would have been executed quite differently or earlier. Even recent schemes could have been carried out more effectively and more cheaply. The value of scientific hydraulic research cannot easily be assessed in terms of money. It renders possible economy in materials and labour while the work is in progress, but the main advantage is that it opens up new vistas of the general physical and economic possibilities of the country as a whole. The investigator charged with the task of examining the *physical potentialities* of a country must take a very broad view and not confine his range of vision to the technique of hydraulics. The knowlegde which is of service to a country may come from unexpected quarters.

To quote an instance: what was the value of the biologist's advice to admit the elvers into the Zuiderzee at night instead of in the daytime? It simply meant the end of the Egyptian plague of mosquitoes and millions of guilders' worth of eels. What was the value in terms of money of the mathematical genius of Professor Lorentz, which enabled him to predict accurately the future tidal levels north of the Zuiderzee dike? It rendered the execution of the project safe, as it enabled us to take precautionary measures against future huge inundations. What again was the value of that inspiration which caused a contractor to recommend the use of the heavy boulder clay abundantly obtainable from the bottom of the Zuiderzee?

Ignorance is expensive and unsatisfactory. True, it is not easy to ban ignorance and to grasp the intricate laws of water and economics, but even the slightest comprehension of the essentials is worth while. Galileo considered it an easier task to study the laws which govern the movements of the planets, despite the astounding distances at which they take place, than to examine the movements of the water flowing under his eyes. But do these movements take place right under our eyes? The main sand movements are right at the bottom, beyond the reach of the human eye.

Every river, every sea inlet, every tidal basin is veritably a sealed book to us until we have learnt to look at the bottom. 'Waterstaat' has realized that the evolution of sound projects demanded thorough research. There had already been engineers with a scientific bent who had studied the ways of Nature, but it was not until 1941 that a research bureau was instituted to investigate the phenomena near the bottom of sea and rivers. It may seem late, but in this field the Netherlands again led the way.

There are four aspects of research with which present-day hydraulic problems can be solved:

(1) Historical research, (3) Mathematical research,
(2) Research 'in nature', (4) Laboratory research.

Each of these fields of research has its individual limitations, as will be clear from the following:

Historical research is based on the so-called film, a collection of sounding charts of a certain area, all on the same scale and to the same level. If the series is continued for a sufficient length of time, say one or two centuries, it affords a clear picture of the regime of natural development of a system of tidal channels.

Research 'in nature' requires the use of different kinds of instruments, such as sand traps, tide meters, echo-sounders, grabs, drills, and such like.

Echo-sounders particularly are instruments of great value, for without them we could not obtain a true picture of the result of the forces operating in a river mouth or sea inlet. This new instrument (1933) is 'the eye on the bottom' for which we hydraulic engineers have longed for so many centuries. The ingenious apparatus known as the Huson, based on radio principles, emits a supersonic signal several times per second to the bottom. There it is reflected and the echo is detected on paper, where it records itself electrically. Although the rate of transmission of sound in water is several times faster than in air, the recordings of the depths thus obtained are accurate to within 3 inches, no matter at what speed a ship is sailing. The instrument also records any unevenness of the sea bottom, reefs, wrecks, even ripples or shoals of fish. This instrument, more than any other, proves that a new day has dawned for hydraulic engineering and that the age of cumbersome measurements with poles and line is dead.

Research 'in nature' is also concerned with prevailing currents and with sand and mud movements – the so-callled sand currents and mud currents. Scientific ally well-founded formulas for their study have been developed and tested by millions of measurements under different circumstances. It is a pleasure to note the numerous variations in the fulfilment of the same basic principles and to

Networks of tidal channels in Holland in which an alternating current (the tide) vibrates. They are being calculated continually because changes by dredging or dammings are planned incessantly.

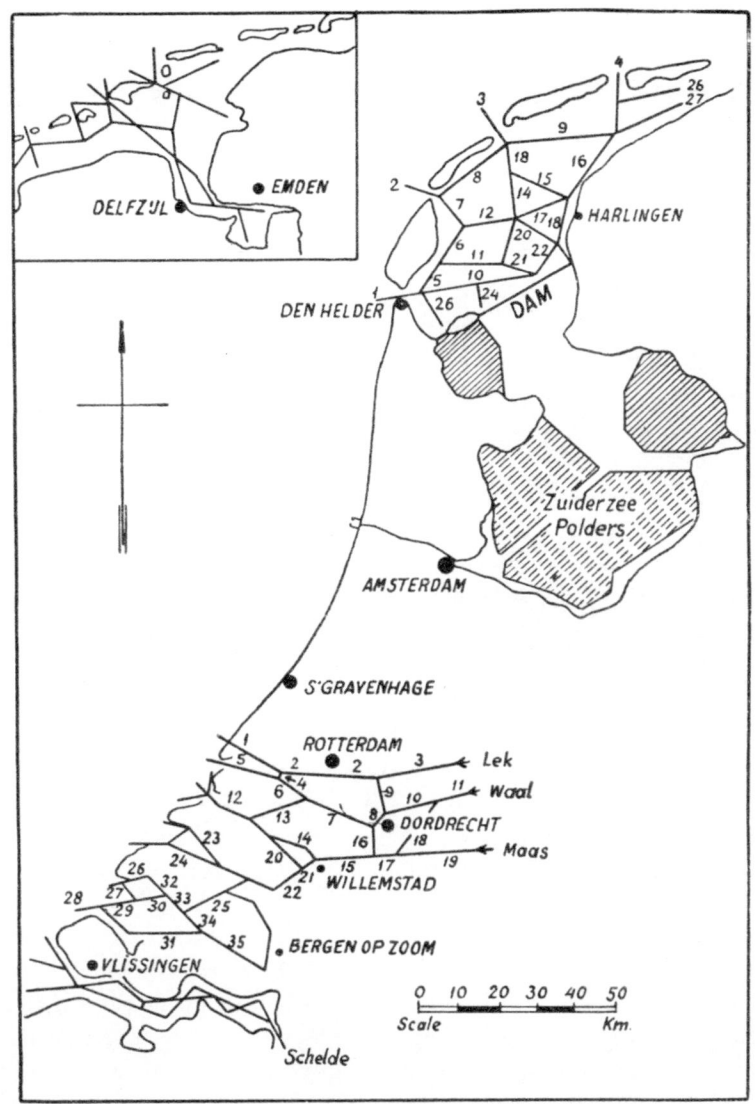

observe that no two sea inlets react in exactly the same way to the same forces. Nature remains an enormous unit in all her diversities.

Mathematical Research dots the i's. It is indispensable for a clear insight into the laws governing currents, tides, sands and silts or salt. The formulae representing tidal movements, however, are extraordinarilly complicated. Often 50 of these complicated formulae have to be solved and the calculations for a single project, employing 15 calculators and modern machines, may take up to 10 or even 15 years. After having developed these formulae as far as possible

143

since 1920 'Waterstaat' is now trying the idea of 'calculating' tides electrically. This is possible, because the mathematical formulae have shown that tidal currents display analogy with electrical alternating currents. In this conception:

direct current	= ordinary river	self-induction	= momentum
alternating current	= tidal current	resistance	= bed resistance
capacity	= tidal fill of basin	conductivity	= conductivity

We found it possible on this basis to imitate Lorentz's tidal formulae electrically with an accuracy to one-half of one per cent. Even considerably more accurate formulae, which take account of the deformations of the tide and which would require a much more complicated mathematical elaboration, can now be solved electrically with also a high precision. A tidal period of 12 hours and 25 minutes is taken as corresponding to an electrical vibration of 1/1000 sec. In the network of tidal rivers, which is represented in the form of copper wires, condensers, resistances and self-inductors, the solution is therefore given 1000 times per second. The height of High Water and Low water, as well as the currents at any point of the rivers of the network, can be measured electrically, or visually shown by means of a cathode ray tube. In this way it is possible to make calculations in advance for any scheme for improving the rivers.

Laboratory Research is a well-established means of approach nowadays to any hydraulic problem. It gains its greatest triumphs in models with scales which are made as large as possible. If the aim is to exercise small cable forces on ships being passed through a lock, or to ensure that a harbour suffers a minimum of trouble from wave action or silting, it is always wise to call in the aid of a hydraulic laboratory. The heights to which waves strike our dikes, the laws governing the movements of sand, the correct streamlining of river banks, these and many other problems can be studied effectively on a miniature scale.

As stated before, however, none of the four branches of research is capable in itself to solve completely and correctly the puzzle of the water and sand movements which will occur after our projects have been completed. Taken as a whole, hydraulic research forms a square pyramid; by collaboration and scientific growth the four sections of research: 1. historical, 2. natural, 3. mathematical and 4. laboratory research, approach each other. They unite at the top of the pyramid and from that point the best project is evolved. When they do not unite, the top has not been reached as yet, and research has to continue.

This four-fold, detailed and scientific research in respect of hydraulic projects is advisable, because such projects usually involve expenditure of millions. A saving of only a few per cent of the original estimate, or the avoidance of a seemingly insignificant mistake, may make a difference of a million in capital

144

A large and well equipped laboratory at Delft, founded and set to work under the able supervision of Prof. J. Th. Thijsse, where all kinds of hydraulic problems – foreign as well as Dutch – are being studied and solved.

expenditure. History proves that we have never tackled our problems with sufficient scientific insight. – Geological, economic, social and planological research, so closely related to most hydraulic projects, are equally, or perhaps more, essential, but these aspects cannot be dealt with here.

Historia docet! See how *inventions* have shaped the Dutch destiny. There are four outstanding historical dates at which the line of development soared higher, receiving a fresh impetus each time because of an invention.

1.   *About 400 B.C.* The beginning of the use of iron in northwestern Europe coincides sufficiently with the start in the **building** of dwelling mounds in the marshes to suggest that the *iron spade* might be the essential invention which made living in the marshes possible.

2.   *About 800–1000 A.C.* The beginning of dike building, especially after the Norman raids ceased, will have followed the invention of *sluices* automatically drawing with ebb, closing with flood. The invention of shipping *locks*, of such great importance for the trade of the Low Countries, followed a few centuries afterwards.

3. *About 1600.* The construction of *windmills* gave a remarkable extension to human power. Land permanently under water could now be reclaimed. Industry soared higher. Emigration of engineering intellect followed. Holland became the centre of the wood mechanical era.

4. *About 1850.* The construction of *steam engines.* This was but a continuation and an extension of the windmill. It meant still greater power. Dredging in sea entrances became possible, etc. England became the first centre of the iron mechanical era.

There was always a lag of time between the actual invention and its full application, a time filled with experiments and infantile diseases. Some windmills were known in the Netherlands in the 14th century, but not before 1600 did their use become general. The first fairly good steam engine worked in England in 1705 and the steam-pumping machines had won their ascendancy over the windmills about 1850; the first good steam dredges came still later.

History teaches that hydraulic research is necessary. Research is the activity no hydraulic engineer can afford to ignore. Laboratory research is quite necessary and also the old saying '*Natura docet*' is valid. Neglecting 'research in nature' because it seems costly is a grave mistake. The true 'cost' of research is revealed by the mistakes come to light after public works have been carried out without the firm basis which only adequate research can give.

The greatest boon of research is that it is aggressive and persevering by nature. It tackles any economic or physical problem and does not stop before its secrets are solved. After this it wants the good project to be carried out. Another fine feature of research is that it may do away with heated arguments of pros and cons, thus saving time and friction. Its arguments have feet and strength in themselves – they can walk alone.

The new great task lies in agriculture. Though the yield of wheat per acre is high in comparison with other nations,

in 1949	Netherlands	1750 kg/acre or 61.2 bushels/acre		
	United Kingdom	1150 ,,	,, 41.9 ,,	,,
	France	780 ,,	,, 28.5 ,,	,,
	United States	405 ,,	,, 14.9 ,,	,,

the output can be greatly increased in an economic way. With us it is not soil erosion which must be combated, but dry seasons and salt. The very large schemes to this end will provide work for engineers for at least two centuries.

20. *Modern Dutch Abroad*

In any second class compartment of a train in Holland you may sit and listen to the interesting recital of the day's destination. One man is going to Amsterdam,

146

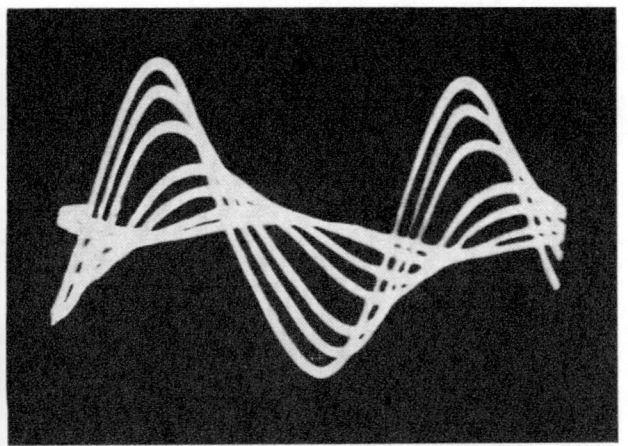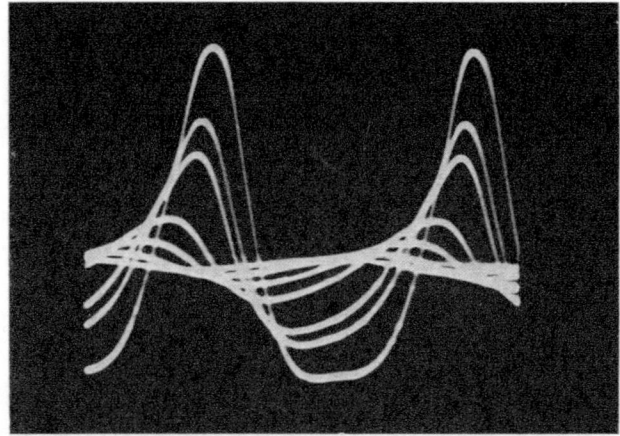

ELECTRICAL IMITATION OF TIDES

The first tidal calculation (of the Zuiderzee enclosure) took 8 years. New schemes led Waterstaat to the use of the analogy between tides and alternating currents. The tidal rise and fall (left) is analogous to voltage, the tidal currents (right) to amperage.

and has a two hours' trip before him, another is off to Haarlem, or further; a third announces triumphantly that he is on his way to Flushing, four hours by train – what a distance. But the silent working man opposite you when he is asked may simply say: 'to Arenas', or to some other unheard-of place. He is a 'polderjongen' and comes from Sliedrecht, or at least from the Alblasserwaard Dike. Going to Australia, South Africa or any other place is all the same to him. He knows his job, and dredging is dredging and mud is mud on all the world's shores. In a few weeks he can teach any man of those shores sufficient Dutch to make a capable assistant in dredging or in constructing brushwood mattresses.

One of the many descendants of Beatrix, Leen Smit, the acknowledged leading firm in the world for towing, who have already towed about 7,000 docks, ships, cranes, etc., plus more than 2,000 items of dredging equipment, and whose losses are less than ½ per cent on their towage of 150,000 miles per year (7 times the circumference of the earth), will look after the mighty dredging tools, and the boss will provide housing for himself and his family. After having heard the song of the dredges on that foreign shore for some years, he will return to Sliedrecht and await the firm's next call to some other shore.

The 'song' of the dredge! Once heard, it is hard to forget. Its mighty groaning and ponderous squeaking is full of contentment and tireless joy of work. You can hear it miles away and it is a sure sign that all is well on board and that much good work is being done. The dredge 'sings' – like a skylark – happily, as long as it is in action. When it stops, the sudden silence is like a calamity – the silence of lost endeavour. Then, as the good-natured machine gets into

147

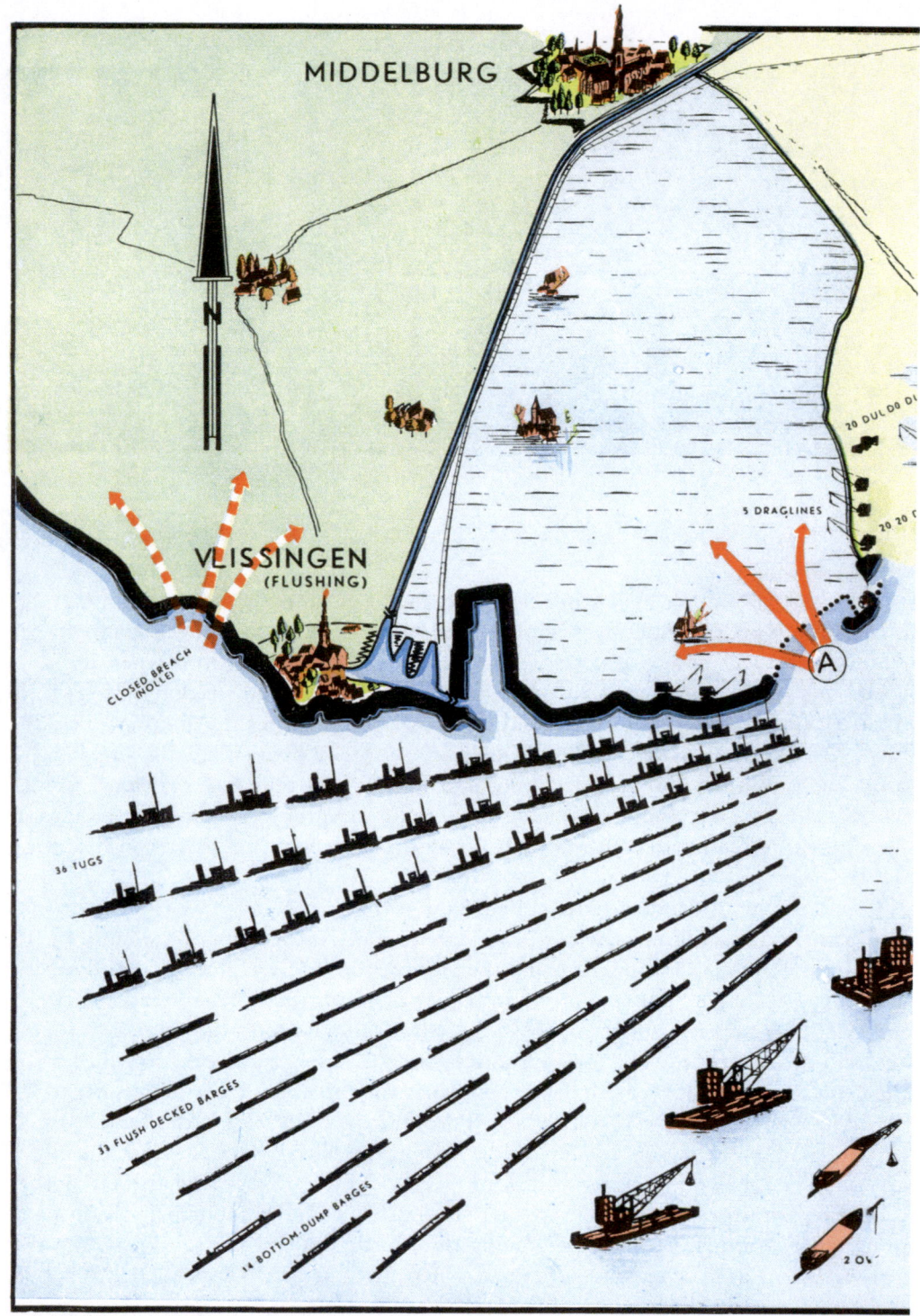

The breach of Rammekens, one of the four gaps the Allies had to make in order to drive the
Germans out of the isle of Walcheren, was closed with great difficulty by using a large fleet of

WALCHEREN
BREACH OF RAMMEKENS

FLOATING MATERIAL AND SUNK ELEMENTS, USED FOR THE CLOSURE OF THE BREACH Ⓐ

SUNK ELEMENTS

27 BEETLES

1 STEEL SHIP

RAMMEKENS

2 CONCRETE SHIPS

2 INTERMEDIATE PONTOONS

2 PHOENIX CAISSONS

FLOATING MATERIAL

3 TRANSIT CRANES

6 SAND PUMP DREDGERS

2 LANDING-CRAFT-TANKS (L.C.T.'s)

47 ELEVATOR BARGES

BAANG BARGES

2 BULLDOZERS

...RTS WITH LOCOMOTIVES

TRUCKS

dredging material, while 34 ships were sunk in the gap. Immediately west of Flushing there was an equally dangerous breach (Nolle).

action again, its peculiar heavy, slow, dissonant chant starts once more, punctuated by the rhythmic or a-rhythmic plop-plop of the heavy substance falling into the barges. There is charm in hearing a dredge at work and the polderman and contractor know that best of all.

If you ask him how and from whom he and his tribe learnt the peculiar trade of making dams and harbours, the man is likely to answer: 'From the beavers', because there were beavers in Holland, long ago. It is possible he is right. The beavers used branches and so do we when we want to dam a river. Their real school was the network of tidal rivers which form the mouths of the Rhine and Maas; especially the intricate mud channels of the 'Bullrush-wilds' or Biesbosch, i.e., the remains of the Great Hollandse Waard, submerged with its 65 villages and 10,000 people in 1421, were their field of action, where generation after generation gathered practice. Here, at the Alblasserwaard Dike, part of which is called Kinderdijk (Baby Dike) sprang up also the factories where the dredges and dredging tools are made.

Most of the best contractors of hydraulic works Holland produced started their careers in the Biesbosch as men trundling their wheelbarrows, or as owners of some small mud barge. These were the practical, staunch and simple self-made men who possessed the rare qualities which go to make an ideal contractor. The great chance for them came with the advent of the steam dredge; millions of cubic yards of sand and clay lay waiting in the rivers and sea entrances of Holland, and any other country, to be removed by the machines. Their forefathers had plied the old-fashioned 'hijsbeugel', the bag on a long pole, with which – straining their muscles to the utmost and wriggling their shoulders and hips – they had brought up load after load of mud or sand. With the turning of the tide their small barge had to be full with mud, as it had to be carried by the tide to the place where it must be brought into some new dike before the next turn of the tide. Not the sun, but the tide governed their work and sleep. They used the tide even when snatching some sleep in the grass on the shore when they fastened the ropes of their barge to their legs so as to be awakened as soon as the tide lifted the barge from its berth, the sign that they could sail. Thus order and method came into their activities. They were perhaps the most hard-working men of all hard-working Holland and the hardiest and cleverest of them could save some money by taking contracts – small ones at first, bigger ones later. After some decades they progressed to the dredge. Not a few of them became men working all over the world.

Patriarchal, simple relations existed on 'The Dike' which is the centre of Dutch dredging families, and this is still more or less the case. One of the greatest contractors, a man who had daily used the spade himself up to his 28th year, and later gradually created a dredging plant which was considered the best in Holland, never found it necessary to make a contract when ordering

150

his dredges to be made. Jan Ten's or Jan Eleven's word – there were so many Jans in the family Smit who made the dredges – was good enough. It was a fine mutual faith which was also maintained towards the Government and towards the working men. Never once in all his long life did this man call in the

Modern cutter and suction dredger of high capacity.

assistance of lawyers, as he wanted to avoid all lawsuits. Difficulties over contracts always were solved by mutual understanding. When carrying out a job which could be easily finished before the winter, he gave orders to proceed more slowly, so that his men should not be out of work. He travelled widely with his dredges and was universally esteemed.

In those first decades of the steam dredges several seemingly small inventions, such as self-emptying barges or sand suction, reduced the price of dredging to one third or one fourth of the original. Organization and technique were continually improved. Progress has not yet come to an end – development continues.

Holland is a family of many children. Some of its sons are engaged in trade,

151

Dredging the new harbour at Capetown.

others in industry. Many of them are tending the 'Dutch garden' i.e., the country itself. This 'garden' is extended by clever manipulation of shallow shores, and it is being made more and more fruitful, so as to make more food available for the ever-increasing family.

Nevertheless many Dutchmen have to earn their living abroad. One of the best things they have to offer is to turn foreign waste lands into well-kept paradises or to open up new territory by constructing harbours or improving the course of a river. They carry the magic wand in their pockets – all they need is the moral and effective support of the country in which they are working. They possess tools, comparatively inexpensive to employ and designed to carry out a sound scheme in the most economical way. Their capability is an inborn instinct, but only by the co-operation and financial support of the country in which they are going to work can their initiative be fully applied.

If there is mutual trust and the Government of the foreign country provides adequate aid, much can be achieved in a short time.

Positive work, particularly when it has a lasting subject, has a charm all its own. A source of welfare and wealth is opened, from which succeeding genera-

152

Sand transport by means of water is one of the most economical ways.

tions may draw freely. The gratitude of those generations is delayed but enduring. The oldest example of such eternal praise is that of the people of China to the Great Shun and the Great Yu. The record tells how in 2300 B.C. floods in China had assumed such proportions that the Emperor Yao decided to entrust Shun, a very intelligent labourer, with the task of carrying out hydraulic works and appointed him as his co-regent and successor. Although he carried out many projects in the course of 30 years, the position continued to be bad, so that once he had ascended the throne, Shun appointed an extraordinary man Yu as his minister. For 62 years Yu carried out his programme; he made a large number of canals and improved many rivers during his long term of office. The greatest of his achievements was to alter the course of the Hwang-Ho in Upper Sjeng-Si and there we still find a stone that records his doings.

Both these men are revered as the saviours of China, because of their positive

153

Dismountable cranes are transported to every place where quays have to be built.

work. It is said that the people of China have observed the rules they prescribed as much as possible and that more than 200 Emperors have carried on their tradition. When even 4000 years ago it was proved possible to achieve much in a period of some 100 years, it is evident that nowadays with our modern and powerful auxiliaries we can do far more in less time.

Labour gives wealth, if it is guided in the right direction. Any Minister or Sovereign who can direct the feet of his people into constructive paths is assured of fame in the future. Mussolini was on the right road when he eventually reclaimed the Pontine Marshes. Also Hitler was doing something to his credit when he ordered the reclamation of the Schleswig tidal sands. Their peoples encouraged them in such positive endeavour, but they continued to jubilate when they were driven into dubious paths. They lacked discrimination.

The Egyptians had their own means of producing such wealth.

'The revenue of Egypt was, in the days of the Pharaoh whom the Moslems call Walid Ibn Masib and who was originally a bankrupt Persian banker and necromancer, upwards of 72 million dinars annually. Of this sum the Pharaoh retained one-fourth for his own use and the expenses of the court, a fourth

154

was employed in strengthening the military defences of the country and *a fourth was spent in improving the land*, constructing and repairing bridges and canals, and in other public works; the remaining fourth was buried, each village receiving back a sum proportionate to the whole of the taxes levied upon it, and the amount so disposed of was allowed to accumulate from year to year until some pressing need or dire calamity should overtake the district.

'Every year at seedtime Pharaoh appointed two officers to travel through the country and inspect the farms. Each of these commissioners took with him a bushel of wheat, and if he found a piece of uncultivated ground sufficiently large to grow this quantity, he sowed it therein and reported the case to Pharaoh, by whom the negligent owner of the soil was at once beheaded. So great was the fertility of Egypt in these days, and so strict the system by which cultivation was enforced, that the commissioners frequently returned without having found a spot on which to dispose their grain.' (E. H. Palmer, 1871).

It is a noble stimulant for any nation to develop its soil and potentialities; to make an endeavour to build up its general prosperity. Nations which have developed their own country have had no cause for regret. One of the highest praises which can be bestowed on a ruler is that he has enriched and embellished his country by great and useful works. This is real glory. It irradiates from the very land.

'None but fools would locate a human settlement in that swamp in northern Illinois, gentlemen. We must refuse the loan these settlers want to develop their village.' This was said about some village in 1830, and this village of fools became Chicago! The faculty to see possibilities started development.

If a nation has eyes to see the *possibilities of developing its country* much can be done by means of the labour at its disposal; home-produced materials can be used, which save not only cost of transport, but also foreign exchange. By applying the principle which contractors describe as *'creating work with work'*, it is possible, for instance, simultaneously to construct a harbour and to use the excavated soil for some fill-in work. The improvement of a river is not a matter of purely navigational importance, but at the same time it provides drainage for the land bordering the river and it may lead to the construction of a hydro-electric power station.

There is a proverb which says that the curse of evil is that it begets evil. Conversely, constructive work causes other useful works to be made and each of them brings riches. When a harbour is constructed, for instance, the logical sequence of events is a system of roads to that harbour and the creation of a town. Industrial plants and general development of the whole country may follow in their wake.

One of the Dutch contractors with a world-wide experience of 35 years, de-

SUCTION DREDGE making a canal. The dredgings are pumped through a floating tube to the dump. *(Photo K.L.M.)*

scribed harbour-making as giving satisfaction to the 'creative instinct' of mankind. Here is his view:

'The construction of a harbour in a foreign country is a really fine task. It appeals to the imagination and to the creative instinct of mankind. There may be more difficult engineering problems which require finer calculation, but harbour construction demands seamanship as well as engineering ability. It means a struggle with the elements as well as the handling of large masses of heterogeneous labour.

'It mingles with and affects the whole town, it is a work which cannot be hidden from view and it becomes a matter of daily interest to everyone.

'When a harbour has to be made on a barren shore, it is astonishing to see how soon a village springs into being and how quickly the place becomes a junction of land and sea routes. Ships arrive with coal, wood and cement, tugs bring dredges, barges and floating cranes, and in a few months the name of the place is known in all the neighbouring towns and ports.

'Not that such a start is easy! New harbours are mostly wanted where there is little or no shelter for ships. Until the breakwaters begin to afford some

156

Special dredging and soil-transporting machine in operation at Donzère, France.   (*Photo I.H.C. Holland*)

protection, the engineers in charge and the crew of the floating equipment live with one eye on the barometer. Salvage operations on sunken or beached ships are a normal part of the job.

'Slowly we get accustomed to the circumstances. Offices and quarters are built, a workshop is made to handle repairs, wells are dug for the water supply, a coal yard has been fenced in and pilferage is reduced to a 'permissible' figure, quarries are opened, railway tracks laid and locomotives take long rows of loaded wagons to the temporary jetties where the rubble is transferred into barges, large mixers turn out concrete to fill the huge moulds, and gigantic cranes are erected to handle the concrete blocks sometimes more than 30 tons in weight. Slowly the local workmen, and those brought from afar, get used to the work. Then the most hectic period belongs to the past.

'At last the whole work is in full swing. Barring stormy days, it is beginning to run as smoothly as a well-managed factory. The members of the staff now know every working man by sight, even the nickname if the real name is unpronounceable.

'The years rush by; there are some red-letter days, for instance when a large

157

reinforced caisson is launched from a slipway and the authorities are invited. The breakwaters reach an imposing length. The children of the staff grow up, the 'Notices to Mariners' become more and more prominent in the Marine Papers and the local overseers gradually assume responsibility. The staff, which has gradually become part of the town's society, gets orders to prepare for the next undertaking.

'At last the great day! The last block is lowered, or the closing caisson is placed, corks pop from champagne bottles, guns fire, and with a wrench the staff leaves a town that has become so familiar.

'Naturally the scene varies with the continent and latitude. In North China where the local labour does not observe Sundays, working hours are long and a Sunday with only seven hours of work is considered a holiday. A severe winter may provide a long rest, however. In South China there are pirates who add excitement to life when they send letters signed with the skull and cross-bones as in boys' story books. Gruesome tales of what happens to their unfortunate captives add some piquancy. I well remember those pirates with their fast craft. Their methods are clever enough and we had real fights even when we had surrounded our working plant with fortifications. Guards kept watch day and night.

'Typhoons are another trouble there. One of our dredges was turned upside down in a typhoon and two sailors could be heard inside the vessel when the storm was over. They lived on some of the pocketed air underneath the upturned flat bottom. It was far from easy to rescue them, because as soon as we made a small hole, the air rushed out through it. By making round drill holes on the circumference of a circle, and putting sticks in each of the drill holes until we could crash the whole circular plate, we were able to drag them into the open, before the dredge sank as a result of losing its air.

'The west coast of Morocco is dangerous because of its waves and rocks, and so are so many other coasts. Especially in Chile, huge rolling masses of water heave up and down to the height of many yards, even when there is no wind at all. This means that there has been a storm somewhere to the west. Standing on a breakwater 40 feet above mean sea level and with a complete absence of wind, one may be washed into the sea by the rollers. A tugboat may be visible one moment and invisible the next in the deep troughs between the billows.

'Arabs are just as careless with dynamite as the Chinese quarry men. At Leixoes in Portugal one of our fourteen divers was attacked by a huge ray, a dangerous fish of about 10 feet square. We hooked that creature.

'As regards local colour, most Dutch water workers can acquire a taste for delicacies like octopuses and strange outlandish stews, washed down with the products of the vineyards.

'Thirty-five years of dredging and harbour-building in China, Tasmania, India, Africa, Europe and America, what a film it makes in my mind, when I come to remember all that happened. I long to see the clear water of the Caribbean once more, though its coral rocks are extremely hard to dredge. Diving is a pleasure there, the water being so clear and warm that the divers need not warm their hands in a bucket of hot water every time they come up for a smoke. And what about Iran? It would be worth while to see whether the Caspian sea has lowered its level still more. The depth of our new harbour decreased very quickly; even while we made it the level dropped 3 feet. It was an unforeseen trick either of Nature or of the Russians who were playing with the water of the Volga, but it should have been known before that the level of that sea could fall so rapidly. The trouble we had when transporting our dredges and cranes through the Russian rivers and canals! The battle with the tropical heat in summer, and with the snow on the mountain passes in winter!

'Thirty-five years and two wars and several revolutions. Our floating plants sunk, our tugs confiscated, our labour scared away, the currency going west, often dropping to a fraction of its value by the time the contract had to be paid – it all is part of this most interesting game of making the world fit for trade and wealth. There is a lure in this kind of work.

'But now the future. The need for harbours after the war is as great as ever. Traffic is increasing and ships must come and go with the minimum of delay. Dredging, fascine work, towage on the high seas, and other Dutch specialities must have some value in times of peace. Having been locked up virtually in a prison for over five years has greatly increased the desire of all of us to work. We want to see our dredges sail the seven seas again.'

21. *The Tools*

When shortly before the 1939–1945 war an order was placed for the extension of the harbour of Capetown, 60 men left Holland. They were accompanied by 6 dredges and other floating equipment.

The dredging fleet which arrived in Capetown was described by the people of South Africa as the Dutch Armada. The task was a considerable one, shifting 11,000,000 cubic yards of soil, of which 900,000 cubic yards were rock. When the dredging in Capetown harbour was finished, the same contractor, in conjunction with local firms, received an order to construct the largest graving dock in the world, one in which even the Queen Elizabeth could be accommodated quite comfortably. This giant graving dock was finished in the record time of 16 months. Its internal dimensions are 1118 × 148 × 45 feet. But greater tasks were undertaken, for instance at Surabaya, with a removal of 105,000,000

EUROPE

1. Verdal
2. Vaerness
3. Esbjerg
4. Hamburg
5. Brunsbüttelkoog
6. Wilhelmshafen
7. Emden
8. Canal Dortmund-Duisburg
9. Karlsruhe
10. Charleroi
11. a. Antwerp
    b. Viersel
    c. Grobbendonk
12. a. Ostende
    b. Oudenburg
13. Boulogne
14. Isle of Grain
15. Sheerness
16. Liverpool
17. Barrow

18. a. Westfield
    b. Leith
19. Dundee
20. Unerick
21. Cork
22. Shoreham
23. Southampton
24. Guernsey
25. Berville
26. Bordeaux
27. El Ferrol
28. Pontevedra
29. Vila Real de Santo Antonio
30. Huelva
31. a. Cadiz
    b. Puerto de Santa Maria
32. Almaria
33. a. Port de Bouc en Berre
    b. La Vera
34. Marseille

SOUTH-AMERICA, AFRICA, ASIA, AND AUSTRALIA

1. Guayaquil
2. Maracaïbo
   (Los Haticos)
3. Rio Catatumbo
4. Lago de Maracaïbo
5. a. Bachagucro
   b. Las Marochas
6. a. La Salina
   b. San Lorenzo
7. a. El Cardon
   b. Las Piedras
   c. Amuay.
8. Puerto Cabello
9. Rio Tuy
10. Rio Chico
11. Rio Apure
12. San Nicolas (Aruba)
13. Willemstad
14. Barbados
15. Sierra Leone
16. Monrovia
17. Buchanan
18. Abidjan
19. Lagos
20. Nana Creek
21. Bonny
22. Port Harcourt
23. Leopoldville
24. Walvis Bay
25. East London
26. Moçambique

27. Mombasa
28. a. Aden
    b. Little Aden
29. Port Sudan
30. Nile
31. Suez
32. Port Saïd
33. Marsa El Bregha
34. La Goulette
35. Bizerta
36. Alsançak
37. Zonguldak
38. Mersin
39. Kishom
40. Kuwait
41. Khärk
42. Bahrein
43. Karachi
44. Bombay
45. Fulta Point (Calcutta)
46. Klong Toi (Bangkok)
47. Sattahip
48. Hongkong
49. Surabaja
50. Freemantle
51. Albany
52. Geelong
53. Melbourne
54. Kurnell
55. New Castle
56. Bluff

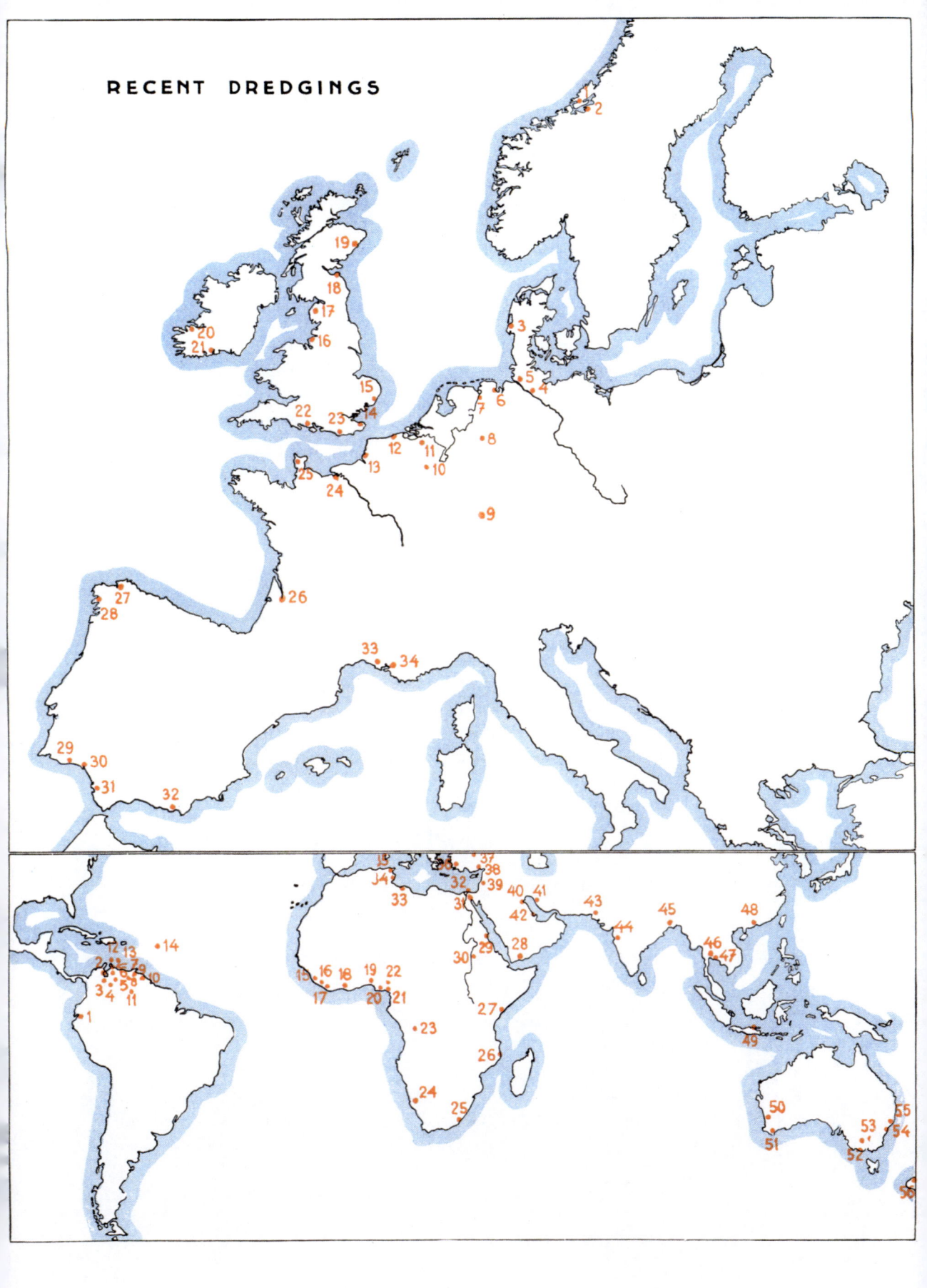

RECENT DREDGINGS

cubic yards, and at Buenos Aires, where also upwards of 100,000,000 cubic yards were dredged.

A single Dutch contractor dredged some 140,000,000 cubic yards in Germany in the pre-Hitlerian era and the total reached by all the Dutch dredgers in Germany was some 260,000,000 cubic yards, so that it is not an overstatement to say that the modern German ports and the North Sea-Baltic canal were, to a very large extent, constructed by Dutchmen.

Before World War II the greatest concentration of dredging materiel ever brought to bear on a project was during the construction of the Zuiderzee dam (1925–1932). The following main floating materiel was on the spot:

11 bucket dredges, 7 suction dredges, 10 boulder clay cranes, 3 boulder clay elevators and 15 stone-cranes, making a total of 46 large items of floating equipment.

In addition there was the auxiliary floating materiel consisting of 215 barges, 77 tugs, 31 houseboats, etc., giving a grand total of 505 vessels engaged on this task.

These figures give some idea of what was necessary to build the great dam. A total of 505 large and small floating craft is a very large dredging fleet indeed and will not easily be surpassed on any other hydraulic work.

The dredge is nowadays by far the most important instrument used by the Dutch to make and improve their country. The technique of dredging has become so good that both a bucket dredge and a suction dredge can make depths up to a 120 feet and more. The transport of solids by means of pipelines can be effected over a distance of about 4 miles. By using intermediate pumping stations this distance can be multiplied.

The entire Dutch fleet of 1960 consists of the following main equipment – minor items not included.:

> 264 bucket dredges
> 4 bucket dredges with suction pipes
> 238 suction dredges
> 22 suction hopper dredges
> 18 elevators
> 252 dump hopper barges
> 296 elevator dump hopper barges
> 722 elevator barges

This capacity is large, when we compare it with the capacity needed for the Zuiderzee works:
264 bucket dredges, or 23 × more than the Zuiderzee plant at maximum,
264 suction dredges, or 38 × ,, ,, ,, ,, ,, ,, ,, .
From this it is obvious that the Zuiderzee works did not put our dredging

fleet under any great strain. As regards the technical equipment, we could have executed several works of that size at the same time. Also it would be possible to dredge a modern Suez Canal every year with this fleet.

22. *Riches*

Looking back along the long lanes of history, surveyed in the foregoing chapter, we find a great desire of the people of the Low Countries to improve their own or any other country; to make good, well-equipped land from bad land, to divert a threat from Nature into a boon.

We saw the 'misera gens' of Pliny's time as ship-wrecked sailors upon their mounds; estimated the incessant spade-work of 2000 years performed under the threat of an endless repetition of flood disasters, and perceived that in the end two strong allies were obtained: first the *windmill*, later the *dredge*. The urge to create fertility and riches will not remain hidden under a bushel, now that modern tools are available. It already found expression in a much quoted sentence: 'Holland is not yet finished. – It will not be finished before the sea is driven out wholly, not before the country has become a paradise'. For this, the generations have striven. The present generation is prepared to work for this ideal harder than ever, partly from tradition, partly from joy in reclamation or any other 'positive' work, and partly from the necessity, imposed by a fast-growing population and a rising sea-level.

Holland may boast some fine cathedrals and town-halls, but, as may be learnt from her history, her real monuments are not in stone but in the local clay. Holland stands as a monument to itself.

It satisfies the Dutch temperament to create economic, sound and useful things. Such things may at the same time be beautiful, but they should serve some useful purpose in the first place. And in the second place they should be neat; neatness being so closely related to usefulness. Rather than pay for the making of beautiful palaces and dazzling official buildings, the farmers worked at the improvement of the dikes and drains, thus benefiting their family estates.

'*Economic is excellent except for diking and draining*' was their ancient saying. And the merchants preferred to increase their mighty fleets in order to strengthen their grasp on world trade. They wrote: 'Cost precedes profits' upon their official building in Amsterdam.

The making and moulding of the country certainly was not initiated by artistic feelings or by the will of potentates, but by the simple desire to live. Our engineering works are not works of pure art, yet some fine qualities went into the making of them, like intelligence, infinite endurance, neatness, cleanness, precision, love of the land and of those who inherit it, the joy of creation and

163

improvement. Cleanliness and great care are reflected by the land. It appears from the air like a cherished, well-kept garden.

The strong creative urge which resulted in the making of this now green and fertile 'garden' has been expressed from early beginnings, and these expressions have not grown stale. Nor can it be said that those of our own times show any sign of degeneration. Here are some old and new sayings – creating a country seems to be an endless joy.

The oldest and most poetical verse of tribute to our activities in the national line is given in the Gudrun Saga, inspired by and dealing with the events which took place on our coast during the Viking invasions. It is about Gudrun's father

> *Hetsel, the Frisian Lord,*
> *Him serve both land and water.*

The ancient bards of 10 centuries ago apparently saw and foresaw our nature quite clearly. Had the Dutch obtained mastery over the water at such an early date?

This ability, apparent from the bard's verse, could be proved in only one way: by demonstrating throughout the centuries that over-population can be conquered to a large extent by those who can make both land and water serve. – Originally people must have thought that the vast swamps, where the Gates of Hell were supposed to be, were already 'full' when perhaps one thousand men hunted in them. Then came the time when 'over-population' caused the making of more than a thousand large hillocks; after this, the 'over-population' of those mounds caused the making of dikes and polders; and later these polders became overpopulated again, so that new lands had to be gained by land-accretion. Also those new lands became full of people again, but the wind could be made to serve as well as the land and the water. Later steam and electricity were added as servants. Now, after so many centuries of constant 'over-population', we have more than 11 million inhabitants, which is more than any other country of the same size can boast of, and this figure is still growing rapidly; – apparently a reason to feel more over-populated than ever. But yet, a healthily growing population means labour with which we can make the most of our possibilities.

And there will always be possibilities so long as there is land and water in the world to serve intelligence and labour. But the 20th century with its wars and troubles will strain our capacities once more. The field must be the world.

Leeghwater expressed another leading thought, cherished now as when he wrote it in 1641:

One of the new farmhouses on the former bottom of the Zuiderzee

If our forbears had been idle
Holland would have gone to nought.
But they struggled like heroes,
And splendid fields are prepared for us.

It is this family-feeling that gives the engineers the idea that they are working for something of long duration and great moment. The realization of their schemes may cost much time, that does not matter – the following generation will go on building. The country cannot be finished in a single century and every generation should take its own share in the improvement.

Yet the projects do not grow all by themselves. Mostly it takes hard work for the 'Architects of the Country', and much hammering. By the patient system of 'here a little, there a little', a fertile bed must be prepared for the seeds of those projects. In a democratic country such major schemes must be discussed and weighed thoroughly by all who are disposed to think about them.

165

Picture taken in the Wieringermeer, the first polder in the Zuiderzee. In 1945 this polder was destroyed by inundation, but soon afterwards it was recovered.

One of the hammerers was Jan Leeghwater when he wrote: 'This great, excellent, glorious and highly laudable, this imperatively necessary work of draining the Haarlemmermeer, must be accomplished'. The man in the street cheered him for saying so, but we had to wait more than 200 years before the Haarlemmermeer was actually drained.

Also Vierlingh in 1570: 'Tidal sands, even those one foot above low water; they cry to be made into a fertile corn field!'

It is not only technique and labour that is needed to make a country blossom like a rose. It also takes diplomacy and perseverance. When swampy regions have been soaked since Adam's time, the people living in or near it are often so accustomed to bad conditions that inertia will not allow any changes. Many people think like the ancient Roman who said: 'Novelties are odious and perilous', and this fear of novelties demands clever handling. Sometimes, or perhaps often, the opposition comes from the top – the providers of money – then strong, invincible arguments are needed.

Action always begets some reaction, not only when stemming or guiding water currents, but also when persuading people to the actual making of great

166

improvements. There is a technique in carrying technical schemes through their infancy and in bearing the brunt of reaction, which reminds us of Vierlingh's advice as to how to deal with the reaction of water. We should not use too much 'fortse' is his advice: we must meet the peculiar reaction of the inert mind with 'sweetness and subtlety'. 'With small bricks you can bouild a castle, but you must *use time* and in an active way have patience.'

The technical part, or carrying out of some scheme, seldom expends our capacities to the full. It is the preparatory work, the organization and general diplomacy required, that absorbs our energy. Clearing away all kinds of obstacles, so that the scheme may flow like a river in its well-corrected bed, is an important part of the work to be done – quite necessary and often thankless. Remember, the scientific facts of public economy win the way.

The building up of the arguments goes on even if a good scheme is obstructed for the moment. The next generation of builders will carry it on. We of the 20th century do not have to worry about this. Recent engineers have found words no less inspired than those of their predecessors of bygone ages, words that are sure to become famous all over the world. They are hewn in a stone standing on the Zuiderzee dam:

A nation that is alive builds for its future.

This means to say that a nation which does not improve its land is not fully alive. Doing something for the future of the country is a sure sign of living, because it is growth. Improving one's land is a duty to the nation, a symptom of natural health. It should be a tradition. The words of one of Holland's foremost contractors, quoted above, will go down the ages. The increase of investment should be at least 2% annually.

When Vierlingh wrote his great work on draining and reclaiming in 1570, he wanted to give it the title 'The Hope of the Poor'. Dredging, draining and reclaiming might rightly be called the hope of the poor countries, because such positive works yield riches. Anyone who has travelled in backward countries and marvelled at the low standard of living there, has perceived a lack of good roads, good drainage, good rivers, good harbours, good houses, good crops, and much else. In those countries former generations did not build enough, they did not put enough into the soil.

The people of those countries may work hard to keep themselves from starving, they may fight their neighbours to the very limit of their energy, but riches come from well-organized and intelligent effort to improve the land and exploit its possibilities. Fortunate is the people whose predecessors developed the country. But what about wars that can so suddenly and completely destroy the works of peace at which so many generations have laboured? – War cannot

167

so easily destroy the works of dredges and other soil transporting machines. But it is true the more artificial a country becomes, the more vulnerable it must be; in Holland hardly a bomb could fall without causing damage to some bridge, lock or house. However, a good farmer ploughs on whatever happens, and a good builder builds on. One follows one's hopes, not one's fears.

Give a hydraulic engineer 30 million pounds and he will create big and useful things: extensive nets of roads, a new waterway, modern harbours, or even a new country where thousands may live. Give the same amount to the Moloch of War and he can make one cruiser from it, – obsolete, if it is not sunk, within 20 years. The creation of the hydraulic engineer may bear fruit as long as people live and work.

What a huge gap between creative work and a destructive machine; between the positive and the negative!

And how modest we country-developing civil engineers are! In 1950 two dams were made in the ancient mouth of the River Maas, the 'immensum os' of the Romans. The largest of these dams is 2000 yards long. This is in accord with our new-old idea: 'Close the coast against the poison and danger of the sea'. A few thousand acres of new land have been reclaimed by doing so and the Maas has become a large fresh-water lake from which the salt-suffering country can draw its water, and on which the holiday-makers of Rotterdam will sail their yachts in great multitudes. The total costs, a lock included, are £ 1,000,000, the economic profits about 4 or 5 times bigger. – All right, but compare these £ 1,000,000 with some luxuries Holland is spending money on. We could make about 7 of such fine economic works for the ice creams which our children are eating in one year, or we could make about 50 of these works for the amount the Dutch put annually in tobacco. Every year, though drunkenness hardly exists in Holland, the consumption of alcohol equals an amount with which about 65 of these big river mouths could be dammed. All the famous huge Multiple Purpose Plans of the Americans could be made in a few decades for only 5 per cent of the ice cream comsumption in the U.S.A. Give the country builders a chance.

Is there any country which cannot be developed into a land of plenty? Once Holland was a poor part of the world – a 'sebstja' or salt-water swamp (a Moorish traveller called it so) – something like the Danube Delta nowadays, or any other fever-stricken morass, habitable only for wild ducks, gulls, geese and mosquitoes. Now it is an efficient and clean country with almost 800 people per square mile. It has the highest longevity of the world; it further has the lowest death rate, the largest production of cereals and potatoes per acre, the highest production of milk per cow and it will soon have the largest national income per square mile.

168

This is almost at the peak of what might be called 'technical civilization', a position which Holland hopes to keep in the future.

There is *everywhere* an abundance of possibilities which call for study and action, so that untold generations may reap the benefits of the riches sown by their ancestors.

CHAPTER IV

A NEW STORM, A NEW START

by Dr. Cassandra[1]

23. *Dutch Floods Abnormal*

Dr. van Veen asked me to add a chapter about the 1953 flood. Though several years have passed, the consequences of this flood continue to demand an excessive amount of energy from the Waterstaat engineers for the future protection of the Netherlands.

Things have changed in the last years. This book was first published after the war to help our civil engineers and dredging contractors who had lost some of their opportunities to serve Indonesia, a work which, before the war, they had done with much gusto and success. The idea then was that their knowledge and skill might remain unsought, but the opposite has proved to be the case. They are working abroad to such an extent that there are hardly enough engineers for the home country now that the flood has struck it.

A flood in the low countries is different from floods in other parts of the world. When an ordinary river breaks its embankments, the flood will spend itself in about two weeks time, and then the water will return into the bed of the river. Again, when a storm flood breaks ordinary sea walls, the salt water, after having done its damage, returns to the sea as soon as the storm is over. Not so in the Netherlands! When its dikes are broken the sea does not show any sign of retreating from the lands below sea level. On the contrary, it stays there, the tides run in and out through the gaps, daily widening them more and more. The sea immediately takes possession of the inundated areas. New estuaries are formed all of a sudden. Creeks are scoured in the once fertile land, marine sands cover that land and fill up its many ditches. The waves destroy the houses, or what is left of them from the first impact. The works of man in a country below sea level are erased very thoroughly, even wholly, once the sea gets in.

When the flood of February 1st, 1953, took 400,000 acres of the best cultural land in the delta area of the southwest of the Netherlands, it was clear that the task of *re-conquering and re-making* that area would be comparable in size to the reclaiming of the new Zuiderzee polders, a total of 550,000 acres. In fact, after we had re-conquered that stricken delta area towards the end of 1953, we found that the average cost per acre was as high as the cost of making a new acre of land in the Zuiderzee. The word 'average' indicates that the worst stricken

---

[1] We must confess now that Dr. Cassandra does not really exist; he was nobody else but Dr. van Veen himself. Introducing this imaginary expert the author got the opportunity to launch his critical remarks about fundamental affairs belonging to the safety against the sea, without coming into conflict with the Waterstaat Authority. *S.*

170

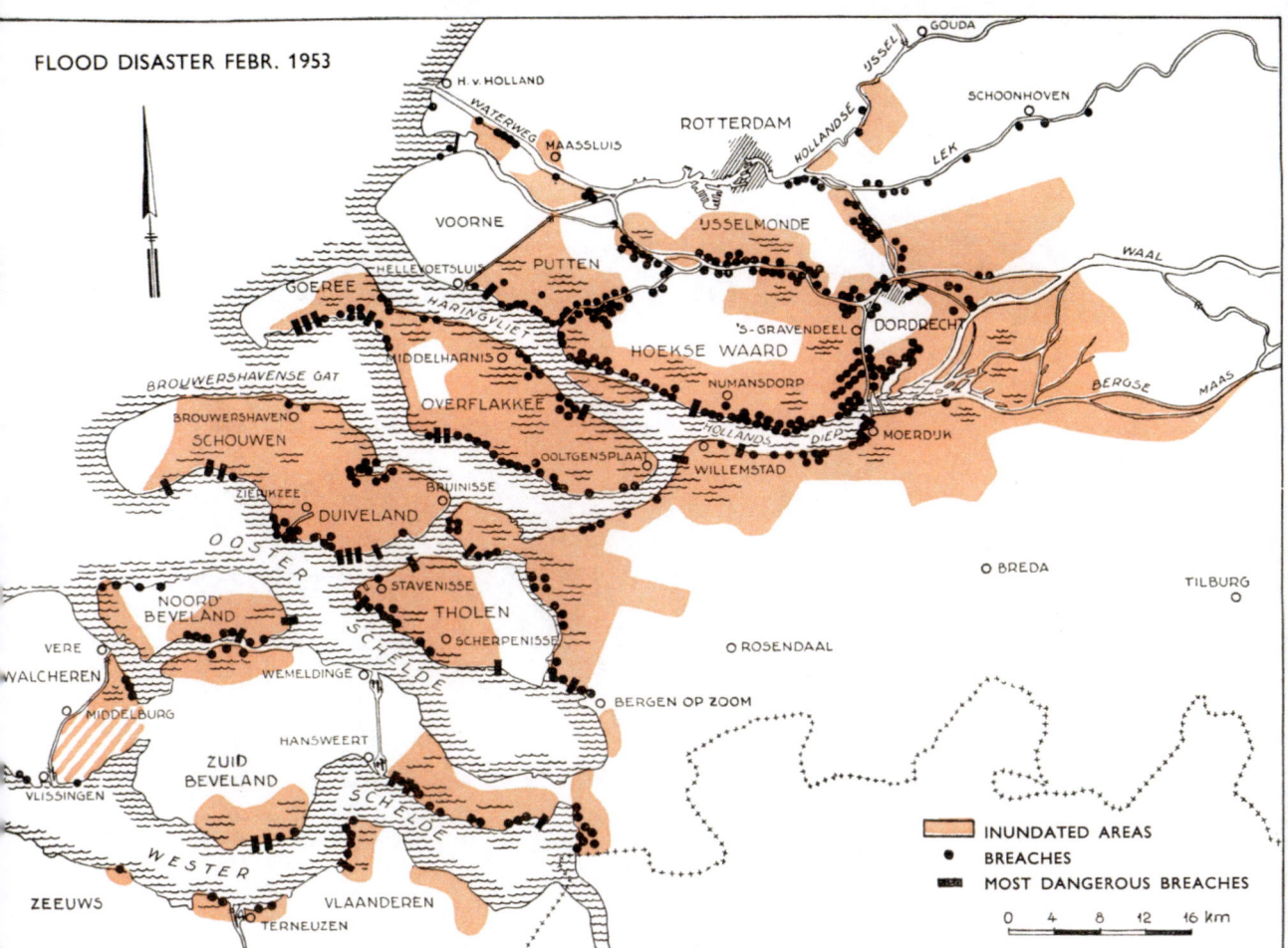

FLOOD DISASTER FEBR. 1953

INUNDATED AREAS
• BREACHES
▬ MOST DANGEROUS BREACHES

0   4   8   12   16 km

The flood of February 1st, 1953, reached great heights and overtopped and destroyed the main dikes in southwest Holland. 400,000 acres were flooded and 750,741 inhabitants were affected. 1835 people were drowned. The 1953 flood, though belonging to the major kind of floods, did not cripple the recuperative powers of Holland. The flood of 1825 inundated 952,000 acres.

islands such as Schouwen required far more for their re-creation, even about five times their selling price per acre. We are not rich enough to permit ourselves such damages.

24. *The Vulnerable Country*

The members of the Diplomatic Corps in The Hague were flown over the stricken district two days after the flood. The diplomats were impressed by the wide, muddy sea. Where were the Zeeland islands? Where were the estuaries between them? All looked much the same. Muddy water everywhere. Indeed, it was like flying over the Zuiderzee.

The help, these diplomats, their governments and their powerful armies and people gave exceeded all expectation, the goodwill expressed was touching; no nation has ever received more immediate or effective help. It was heartening to know of the great sympathy of our Queen and her House for those who had suffered in the flood. It was heartening to have Her encouragement. And

171

Ebb-current through a breach in a dike, February 1953. Tidal gullies cause small waterfalls, which erode backwards. Those gullies thus lengthen, widen and deepen daily, the sand in the soil is being swept either into the sea or is deposited on the inundated lands. The storm of 1953 created 76 of such stream-carrying breaches.

heartening too to see a world-in-unity sending an abundance of aid, including the most modern equipment, such as helicopters and amphibian craft. Goethe wrote of such a moment of Faustic delight: 'Gemeindrang eilt die Lücke zu verschliessen' and he would say to it: 'Stay, I pray, thou art so grand'.

The accompanying chief-engineer[1] told the diplomats that Waterstaat had some hopes of having the broken dikes repaired before the next winter. Hectic and uncertain as the work would be, some inundated parts would be saved within the next months, he said. Here a voice from the audience was heard: 'Impossible! Not months, years! It will take years!'

Was it preposterous to hope for the reclaiming of 400,000 acres in one summer, when the Zuiderzee works with its 550,000 acres were taking 50 years? On the way home, in his car, one of the Ambassadors expressed this view to the engineer; better not voice all hopes.

The latter weighed the arguments silently: The dredging fleet of the country was no longer hampered by the aftermath of the war as at the time when Walcheren had to be re-conquered; the Zuiderzee works could be stopped for the time being and the equipment could be sent to the southwest; spring and summer were ahead, a second storm was therefore unlikely to occur. Direct action

[1] Being Dr. Johan van Veen himself. Nobody could better give information during these flights. *S.*

172

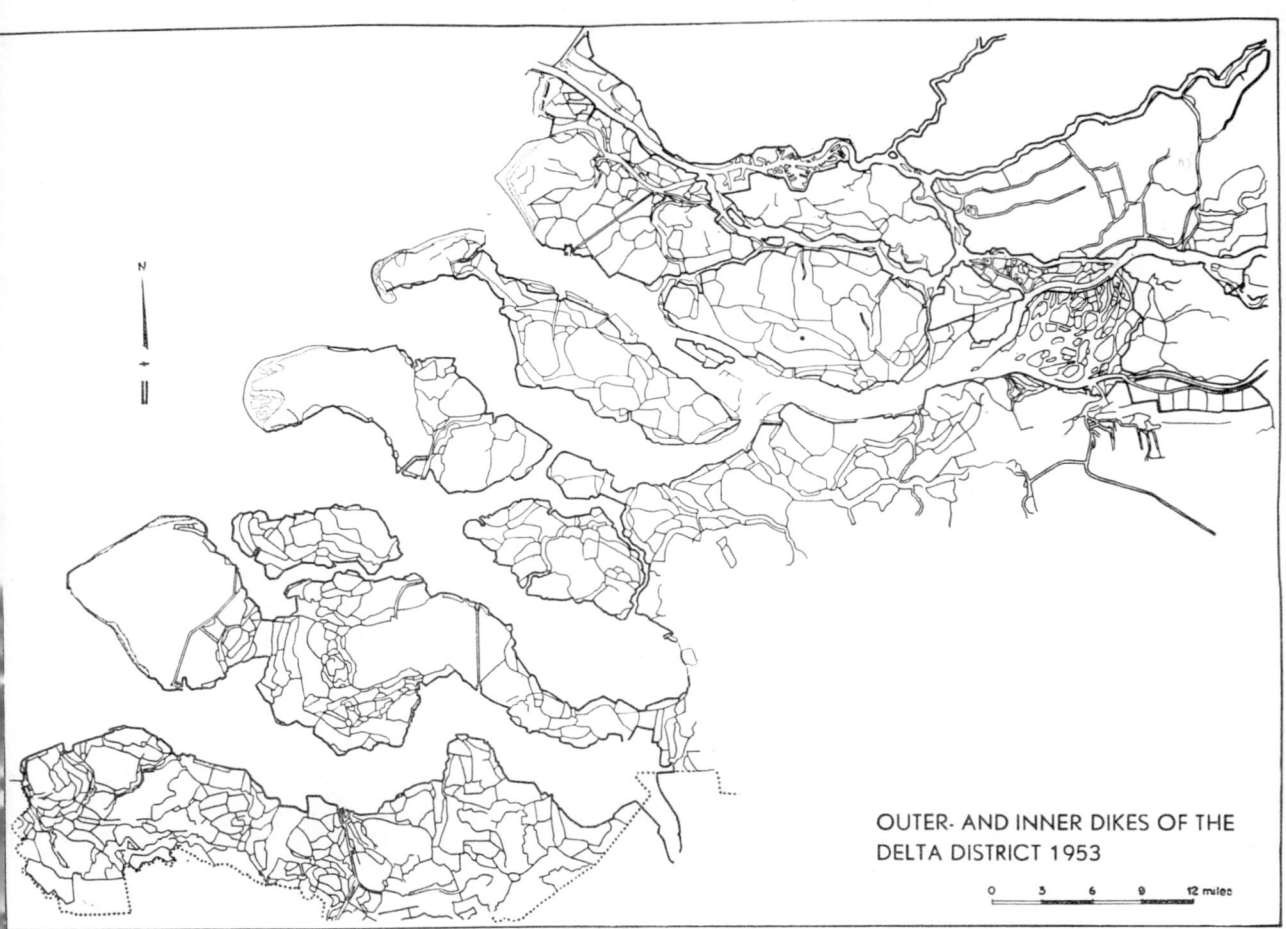

<figure>
N

OUTER- AND INNER DIKES OF THE
DELTA DISTRICT 1953

0   3   6   9   12 miles
</figure>

Zeeland's struggle to rise above the water has been hard indeed. Its ancient device 'Luctor et Emergo' and its enormous amount of inner dikes prove it. Each dike, still existent, was made to extend the islands bit by bit. Outer as well as inner dikes were broken by the 1953 flood.

to close the gaps had been taken by the local Boards and by the army, and had already led to remarkable successes within a mere two days. Not all of the inundated lands were so low. But many stream-carrying tidal breaches had to be closed (later it transpired that there were 89 of them); how wide were they and how powerful were their streams? Many hundreds of miles of dikes had to be repaired. A single summer was a very short time. Yes, he might have suppressed his hopes. But no! Waterstaat was planning to close in the future all the large estuaries of the Zeeland coast, it should therefore be able to close the relatively small inundation gaps in a short time. If not in this summer, the task would become more onerous, and therefore it was certain that every possible exertion would be made by all concerned. Anyhow, time would show quickly enough whether the nation and its engineers would be equal to their tasks. Heavy tasks as had not been taken up in the world before.

He mused further. The great work of closing the Dutch coast, the plan for which he and his men had worked so hard during so many years, would surely

173

Breach at Ouderkerk. 1953. At the moment this picture was taken, the level of the river had dropped about 10 feet.

start, now that the existing dikes had proved to be inadequate. The closing of these terrible storm-breaches would not be more than a mere beginning, just a prelude. The repair of the broken dikes would be undertaken so vigorously (he was sure about that), and successfully (he hoped), that the nation and its engineers might be inspired with necessary courage to commence the task of closing the whole Dutch coast, thus making it almost invulnerable for a long time to come.

How vulnerable the country was, how utterly necessary it was to make one short but very strong line of defence right along the North Sea! His thoughts went back to that fatal Sunday morning. The radio had boomed: 'Many dikes have been broken. All soldiers on leave have to return immediately. There is especial danger at Ouderkerk, there is a wide breach at Ouderkerk.'

Even the dike at Ouderkerk, protecting Central Holland, had been broken. The lowest lands of the country and the main towns were in danger! He knew these dikes very well; they were the weakest in the country, floating on a layer of soft peat which had been settling slowly; they had required heightening many times since they had first been constructed about 1210. Hundreds of houses had been built upon their sides, the very walls and windows of many of these houses

174

Breach at Vosmeer, one of the many, many breaches caused by the flood of February 1st, 1953. Vosmeer is the place where the Roosevelt family came from.

had to help prevent the flood from inundating the millions of people living in Central Holland. Unreliable and too low, wholly inadequate, were these old dikes!

Vivid visions and voices of the first day! The Zeeland islands were lost anyhow, Central Holland might be saved if immediate action would be taken. Therefore off to Ouderkerk in the storm immediately. Over the road on the top of the dike, or what was left of it. The car was difficult to handle. All able-bodied men were already at work, using the utmost of their strength, not speaking a word. The student engineer who came to the halted car: 'We worked frantically the whole night. You warned us at Delft a week ago that our dikes were in a dangerous condition, do you remember?' Off to Ouderkerk! The storm took the water out of the river and dashed it with gusts over the dike and over the houses built on its sides, it was just like dense, horizontal rain drifts. Here was the Niagara, 50 yards wide, roaring above the noise of the wind, falling into the

175

The breach of the dike at Schelphoek, Schouwen (seen at the left) was closed by means of a ringdike, 2.65 miles long. 235 concrete boxes, 36 × 23 × 23 ft. and 1 Mulberry pontoon of 7000 tons were used plus 2,200,200 sq.ft. of willow mattresses.

low polder, already inundated. Large blocks of peat had been washed out of the sub-soil and had stranded further inland. How to strangle that waterfall? It had looked so utterly impossible to subdue it. A church at the edge of it had partly collapsed.

Where to get help? The telephone, telegraph and road connections had been severed. The village on the other side of the gap could not be reached. There was the road to Gouda on the nearly-destroyed dike, but it was almost impassable and blocked by farmers with their cattle. – The little, weird farmer in his clogs, red eyed with cold and lack of sleep, crooked and stocky as an old pollard willow, had come stiffly very near. He looked the engineer full in the face, earnestly but not saying a word, then shaking his head several times, with a question in his eyes. Plucky fellow that farmer, who refused to open his mouth, suppressing his fears. Above the storm the voice of a young man: 'Sir, I have a bike, shall I go and find some ships to be sunk in the gap?' 'By all means do, be quick'. Another

176

voice, beseeching: 'Sir, we have heard that you wanted to blast our church on the dike to use its stones. Please, could you not spare our church?' Yes, the 'Niagara' had been closed that same Sunday at 4 o'clock, and the church was still standing. The soldiers arrived about noon when the water in the river had subsided several feet. How they worked! How the farmers, students and everyone worked!

How strange that Central Holland should have been saved while the stronger dikes of Zeeland had succumbed. Incredible things might happen, but here was a rare wonder. At so many places the dikes and sluices of Central Holland had been on the verge of breaking. Voices again: 'Can you tell me why we, the farmers and townspeople of Central Holland have not been drowned, whereas those of Zeeland have?' Answer: 'Sheer undeserved luck!' Visions of dilapidated ancient sluices, with their broken doors, broken masonry and broken floors, still standing. The harmonium on the table in the farm so that the cherished instrument should not become wet. 'At what time did the first water run over the dike at Ouderkerk?' Answer: 'I know exactly, my daughter heard it here in the room above the noise of the storm. She said: 'Hear, father, the water is now rushing over the top of the dike.' It was half past one then, the time for lowwater. We expected a rise of another 7 feet by then, but the water did not rise that much. The dike broke at five this morning. We had all taken refuge in the church'.

Yes, something had to be done if Holland-Proper wanted to survive another flood. The re-conquering of the inundated Delta districts would be the prelude to prove our mastery and fitness for bigger works. It would be a test to find out what we could do in one summer. The real works were to follow, works to ensure that even Central Holland with its five millions of people, its 75% of importance to the Dutch nation, and its very low lands would be safe for another thousand years to come.

25. *Great Safety Projects Ready*

During the summer of 1940 the Dutch discussed among themselves the character of John Bull. 'His nose must be punched first',we said, 'else he will not fight well. After the fight (nobody doubted the outcome), he will slap himself on the chest and say: good old England, and after that pursue the interesting and profitable paths of peace.' In 1953 we resembled that character. Now our own face had been punched, and we too had defended ourselves vigorously. There was the danger of self-satisfaction after the great exertion and its successes during the summer of 1953, but luckily, the nation and its engineers did not rest on their laurels after closing the 67 stream-carrying breaches and more than 500 less dangerous ones, in one summer. The newly-installed Delta Committee stressed the fact that the great work of closing the southwestern estuaries must start as soon as possible. This, it said, was a dire necessity.

There has been satisfaction with the prowess of the vast body of men who fought at the gaps as there has been gratitude for so much international help, but by 1954 the struggle with the streams, which had been maintained day and night for 10 months had become history. The works to come were so much larger and so much more difficult to tackle. All available energy was in the preparation of these future works. Indeed, there was no time to look back and feel self-satisfied. In fact the construction of the Delta Works has not been postponed, not even for a single day. The building of the storm flood defence some miles east of Rotterdam started in January 1954.

When, in 1667, Hendrik Stevin wrote about the problem 'How to get rid of the fury and the poison (the floods and the salt) of the North Sea', he added: 'Ha, people will blame me, they will compare me to a prancing horse', but he maintained that, in some faraway day, some generation would actually shut the Dutch coast against the 'fury and the poison' of the sea by making dams from island to island.

The generation of engineers, between the wars, who had discovered and read these words then, had had no desire to be compared to a mettlesome horse. Stevin's soaring imagination was no better than Jules Verne's, they thought. Their standpoint was a practical one, less imaginative perhaps, but open to research. They started in 1929 a general study of our tidal waters, not knowing where this study would lead. Many men and ships were engaged in this investigation. In those early years scientific measurements were made in the estuaries and even in the Straits of Dover, that 'beginning' of the Dutch sweep of sandy coast. The driving force behind the idea had been to get more knowledge in the belief that knowledge leads to worth-while discovery. The genesis of the big schemes had been nothing but a simple belief in research.

After some initial groping more and more light was shed on the subject, as the study led to the discovery of the exact laws of the tides in the Zeeland waters and of their sand movements, to the knowledge of the existing but hitherto almost unnoticed salt and drought damage in the lowlands, and also to the awakening of our minds to the imminent danger we were in. Research is the key to the chest where many wonders are stored. Great possibilities, far beyond expectation, were found for improving the safety and the fertility of the country. It could be shown that here was not only 'periculum in mora', but also huge damage.[1]

After these discoveries the war came. No chance for much direct action then,

[1] It may be stated here that Dr. Johan van Veen himself has been the soul of the extensive studies in the years between 1929 and 1953. He was a real master of the floods and in fact he must be considered as the 'father' of the Delta Plan.   S.

Electronic computer based on the analogy of tides and alternating currents. Every 'tide' takes one thousandth of a second.

but 'la vérité était en marche', and nothing could stop it any more. Only the wish: let no flood appear while we are in bonds. Some dikes were improved while the war was on. Also after the war, action was greatly hampered as long as the nation's economy was disrupted. Study and research went on at top speed, however, because it was evident that the 700 miles of dikes in the southwest would not be able to ward off the calamity, already overdue. We would not heighten all these dilapidated dikes, it was even impossible to do so. Some great scheme had to be developed as quickly as possible to cure all the ills of the deltaic country in the southwest.

Nearly 600 reports have been written about the existing conditions and the possible plans for the Delta districts, while about 500 tidal calculations for as many various plans were made; also about 500 model tests. This took more than two million man-hours. The calculations were amongst the most difficult imaginable. Each calculation involved the solution of about 50 differential equations as an average.

179

The stage of considering seriously the closing of the Dutch coast was reached ever so reluctantly and only step by step, Stevin's idea being so colossal and fantastic – and so expensive. The engineers slowly came to realize that there might be a possibility of closing large tidal inlets with sand bottoms if our technique would advance a few steps further. They even drew, hesitatingly, some lines on the maps; and consequently received the name of 'line drawers'. Would our postwar technique really be advanced enough to close the Haringvliet near its mouth, a channel 3 miles wide and 50 feet deep? Could 10 sluices be made there, together 3300 feet wide, for the water and ice of the Rhine? Would the North Sea waves permit our men to work? By December 1925 the engineers of Waterstaat felt sure that the Haringvliet could be closed though it would be risky.

The new Minister of Waterstaat Mr. *Algera*, dispelled the technical timidity of the engineers as soon as he took up his office, ordering them to fix their thoughts on the closing of all southwestern inlets between the West Scheldt aud the Rotterdam Waterway, the largest and deepest estuaries in the whole country and all interconnected by large channels behind the Zeeland islands. The Haringvliet, for which the engineers had dared to make preliminary closing plans, was the smallest of them except of course the Veerse Gat between the isles of Walcheren and Noord-Beveland.

The Minister's order was dated December 2nd, 1952. The engineers adored such a Minister! Then they sat down to more calculations, those terrific and almost unending tidal computations.[1]

Exactly two months later, as if it had heard the order of the Minister, the sea, the foe which the engineers were trying to outwit, made its sudden attack on the area of their fears and hopes. The estuaries which we were ordered to strangle, proved their strength. One blow in a single night, and almost the whole deltaic region was wiped off the map. The low islands had become so many tidal basins. The nation shuddered when it realized the incredibly narrow escape of the central provinces with their large towns and about five million people.

This was the provocation! We had received our blow, only we had not expected that blow to come so soon and suddenly, nor to be so painful. Still, we must see that it was only a mild blow, compared to the one we had so narrowly escaped. Our generation had not had much experience with storms, our history had not been studied enough. We discovered, or came to feel to our deep surprise, that our opponent the sea was really a relentless foe, who actually could and would annihilate the main parts of the Netherlands. As children we had been told in the schools that we had a dragon living next door, and of course

---

[1] It is almost incredible that the first Delta Plan could be developed in this short period of some weeks only. It was dated January 2 th, 1953! *S.*

180

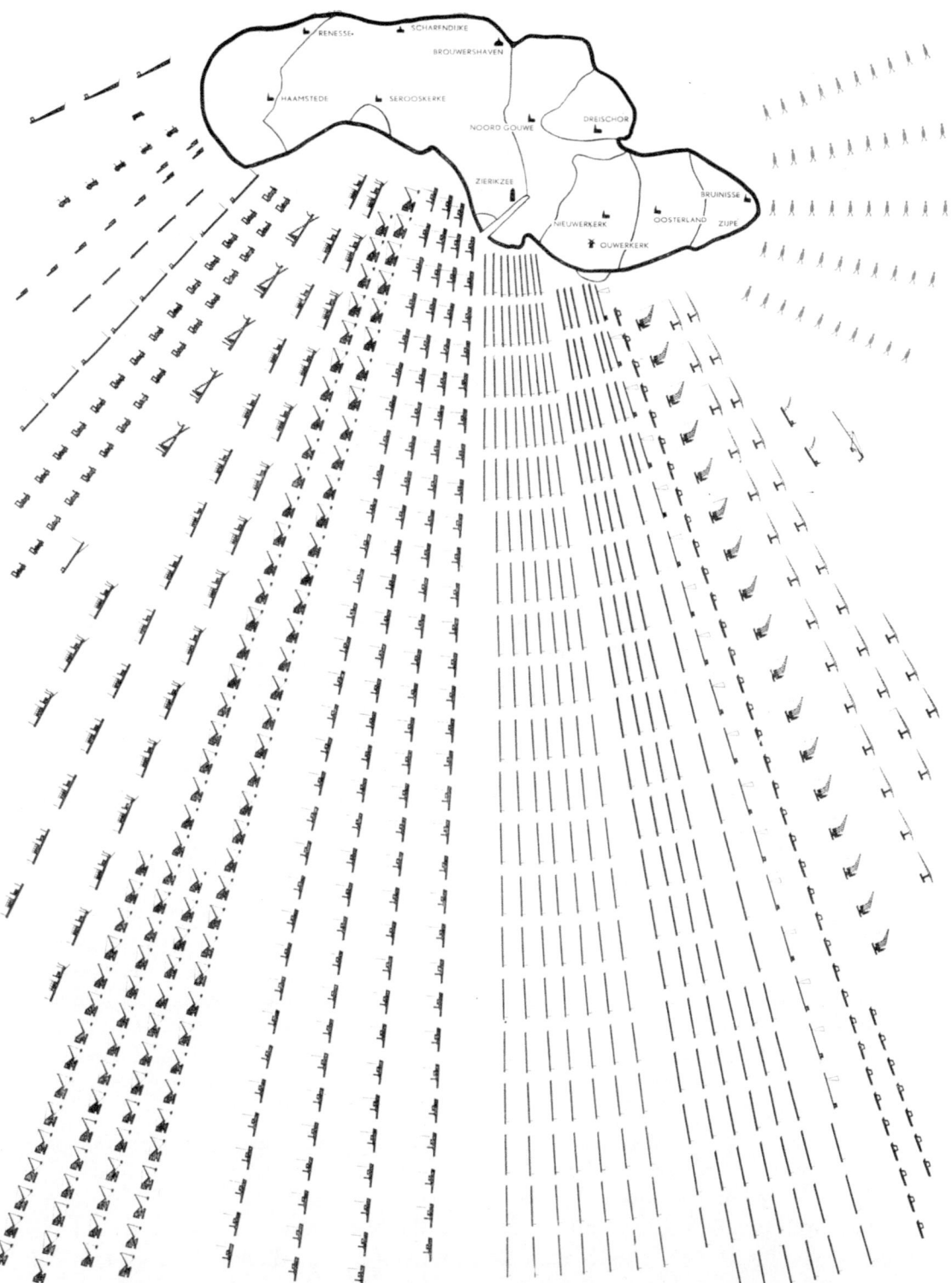

The closing of the breaches in the coastline of Schouwen-Duiveland took a gigantic fleet of dredging and construction material; 126 tugboats, 283 barges of 300 to 700 tons, 38 dredges, 49 cranes, 105 draglines, 43 launches, 53 locomotives, dukw's, bulldozers, and so on. About 5000 workers have been employed here during several months.

we knew it was there, but we had never ourselves seen and felt the quick, sudden, tiger-like stroke of that dragon.

The fight is now on! We are very far from feeling victorious. The foe is still unbeaten and it is able to give us another blow with its paw, even a worse one. We cannot slap our chests as yet, because it is evident that there is a crusade before us of long duration and of great exertion. But the great projects for safety and fertility are now in full swing. The 'final' victory over the sea in these regions may be expected between 1975 and 1978. Of course there never will be a real final victory over the sea. Our struggle will go on for ever. We humans may strive for a very high degree of safety but absolute safety against flooding is unattainable.

Meteorologists, mathematicians, hydrographers, economists and many other scientific experts are helping to make the project a multiple-purpose success. It must not fail. It must be an El Alamein victory, long prepared but successful at the first stroke.

26. *Engineers and Fleet on the Alert*

According to the present projects, eight dams will be made, four of which will lie in the mouths of the main estuaries. The total length of the dams will be about 25 miles, the depths in the mouths varying from about 30 to about 100 feet. The sand at the bottoms is the loosest imaginable. Boulder clay, which played such an important part in closing the Zuiderzee is not to be found in the neighbourhood. Stones have to be bought from our neighbours. Although it seemed to be a ridiculous idea to introduce the shaly substance of the Dutch coalmines in hydraulic engineering, the experiments have proved already the great worth of this revolutionary idea. The work will have to be executed as quickly as possible, because the tides, storms and rent-losses will not allow it otherwise. The estimate for the whole work is 20 to 25 years, which seems rather short. The Dutch Government has decided by special law to carry out this scheme. The works in the Delta-area were started in the beginning of 1954, some years before the Delta Law was discussed in the Houses of Parliament. Everyone considering the closing a number of sea arms in the southwest region as a matter of life and death for both nation and country, the Minister of Waterstaat was endorsed to start these works as soon as possible. Not a year should be lost, not even a month or a day on the way to greater safety.

The Government was faced with the certainty that something must be done.

There were two quite different possibilities. First, heightening hundreds of miles of ancient and unreliable dikes, which in the course of several centuries have been broken at thousands of different places and always repaired in a hurry. Secondly to make the short, strong line of defence wholly reliable and of modern construction.

182

Another reason why action was imperative is that the whole country is subsiding, sinking gradually lower and lower. The geologists tell us that it has sunk 7000 yards in several millions of years. Moreover it is to be expected that the sea will continue to rise some 200 feet in consequence of the melting of the ice in Greenland and in the Antarctic. This rising of the sea proceeds slowly, but in the course of 72 centuries it has risen 55 feet, according to the reliable C-14 investigations of today, based on the radioactivity of peat layers in the soil. Some effective measures must therefore be taken, if not by us, then by our descendants.

There are also the sand losses of the estuaries, which make them deeper and their currents more threatening every year. Submarine landslides occur year in, year out. The upkeep of the coasts of the archipelago costs enormous sums.

Further, crops could be improved if more fresh water were available. The Zeeland islands badly need good roads to the mainland, roads without ferries. The new dams will have roads.

Finally, there was the argument that we have an army of specially experienced water fighters and that there will be not much chance of any future generation getting more experience. We therefore we could not allow time to pass without direct action. We have the men now, and we have the tools, so why delay? Both have proved their efficacy. The dangers which arise mainly from the country sinking slowly under sea level, *must* be faced. There is no hope of escape by minor remedies.

Stevin, in 1667, had in mind that the salvation of the country would come from some future generation. We seem to be that generation. Our engineers have had particular experience in such work. They have followed a special course as it were, in closing tidal inlets. The former generation had no such training, because no heavy storm occurred in the last century, except, of course, the January storm flood of 1916 in the Zuiderzee district. It is not very likely that future generations will ever have such training as may be seen from the five main tasks which the present generation of engineers have been given in their 'school':

1. Closing the Zuiderzee, 1927–1932. An expensive and risky work, already exceeding common engineering experience, it is true, but the tides were only about 5 feet, and the depths no more than 15 to 50 feet. Moreover, there was boulder-clay available.

2. Closing of the four war breaches in the dikes of the island of Walcheren in 1945/46. Though the island was only 50,000 acres, this was a strenuous fight. We barely managed to do it, even with the experience obtained in the Zuiderzee. The main difficulty was caused by the higher tides, 12 feet on an average.

3. Closing of the Brielse Maas, one of the four mouths of the Rhine, 1950. The sinking of many ships into the tidal maelstroms of the Walcheren breaches was now transformed into a new method of smoothly and quickly closing the

last opening by means of a concrete 'door' of nearly 3000 tons at the turning of the tide.

4. Closing of the Brakman, an inlet on the Western Scheldt, 1952. The tide was high there. Two 'doors' of 7000 tons each were necessary. The work nearly failed, as an opening between the two pontoons was not wholly shut. This should not have happened. A fight of a fortnight's duration was necessary to shut this last small hole, which threatened to ruin the entire work.

5. Closing of 89 tidal gaps, in the summer of 1953. The largest of them, Schelphoek, was 580 yards wide, depth about 120 feet. We passed this last exam with flying colours. Nevertheless the engineers had to confess that they had reached in this operation the limits of technical possibilities at that time. For further and bigger operations new technical development was absolutely necessary. Now, nine years after the flood disaster, we may say that the capital problems to build and to close the dams in the sea arms are solved in a brilliant way.

This crescendo gives us hope. When we compare the fleet of Walcheren (page 148) with the fleet of Schouwen-Duiveland, we see a marked increase in size. There are only eight years of schooling between Walcheren and Schouwen, but the increase in the plant employed is striking. Holland may be safe within a few decades because of the lessons provided by these crises and necessary works. The fleet and its men are in good shape.

The compulsory schooling in closing tidal inlets was not for the civil engineers only. During the eight years, experience was gained progressively by the engineers as well as by the captains of the tugboats and barges, by the mattress makers, the truck drivers, the cranemen, and all the other experts. They were all specialists and were further specialized. Though they already knew how to work together without talking, their skill and co-operation have been very much perfected during the last 25 years. There is no time for talking when a gap has to be closed in the few minutes' time between the tides. There may be a leader, using his hands and arms like the conductor of an orchestra, but everybody must know exactly what to do at the right moment, and also what the other men are going to do in the next moment. When after the closing, the whole work threatens to be swept away by the new incoming tide, and the white foam spouts from underneath the still rickety structures, the initiative of all the alert watching men must save the new dam.

In 1953 only a few of the 89 closings were unsuccessful. In such a case several concrete blocks of some hundred tons sank into holes scoured by the swirling waters, and the work had to start anew at a different place. The second closing of the last of the gaps took 10 weeks during day and night after the first closing of that gap had failed. It had to be closed with four large Phoenix pontoons, of 7700 tons each, and it had to be done at midnight, being the only favourable moment between the tides.

Economy may raise its mighty voice here: 'Have you been spending more money on the re-conquering and re-making of the island of Schouwen than that island is worth? How much? About 5 times more than its value? Did you never think about abandoning that island? How many times after each flood in the past have you 're-conquered and re-made' your lowlands?'

We have not yet heard such words spoken by a Dutchman in a serious manner, and we do not expect to hear them, but British colleagues hinted in this direction in their polite manner: Did not the Dutch engineers 'show too much 'heart' in such technical matters?' They thought we did. What about the enormous stone defences on the coasts of Holland, if compared with the cheap, wooden groynes of Great Britain? Waste of money? Too anxious about losing some inches of land?

It seems to be difficult to see the troublous technical problems of an abnormal country like Holland. The sea defences of a country with cliff coasts and salient capes must be different to ours. There are no capes or cliffs on the Netherlands coast. British shores can be allowed to recede; receding they provide much sand and pebbles, which is a coastal defence in itself. The low Dutch coast cannot be allowed to recede; receding it gives no pebbles and hardly any sand. It needs capes, but these must be made by men. We have scores of strong artificial capes, which, from the air, look like so many spears to ward off the threats of the enemy. There is no direct economic value on the credit side of these constructions; they only cost money year in, year out. But they are a necessity. The question of the economy of diking seems to be an age-long question, asked many times of the dikemasters of bygone centuries. The answer these unsentimental men gave was available in an old proverb, put into 'words simple as grass':

> '*Economy is good, except for making dams and dikes*'.

It is the wisdom of folk proverbs that makes them survive:

> *No one knows who said them first, these small bits of wisdom,*
> *Clear like spring-water, tangy as wood-smoke,*
> *Heavy with meaning as stones.*   (*Cairns*)

Long-term economy differs from short-term economy; state-economy from private economy. Slowly and tardily the technocratic mind finds its way into the realms of *public-economy*, based on full employment, and *super-economy*, based on safety, health, preservation and extension of life, and other imponderable items. Necessary wars – the Dutch struggle against the sea is such a war – ask for higher reasoning than mere short-term economy can provide.

Homo-economicus must not turn arguments upside down. Economy is the

185

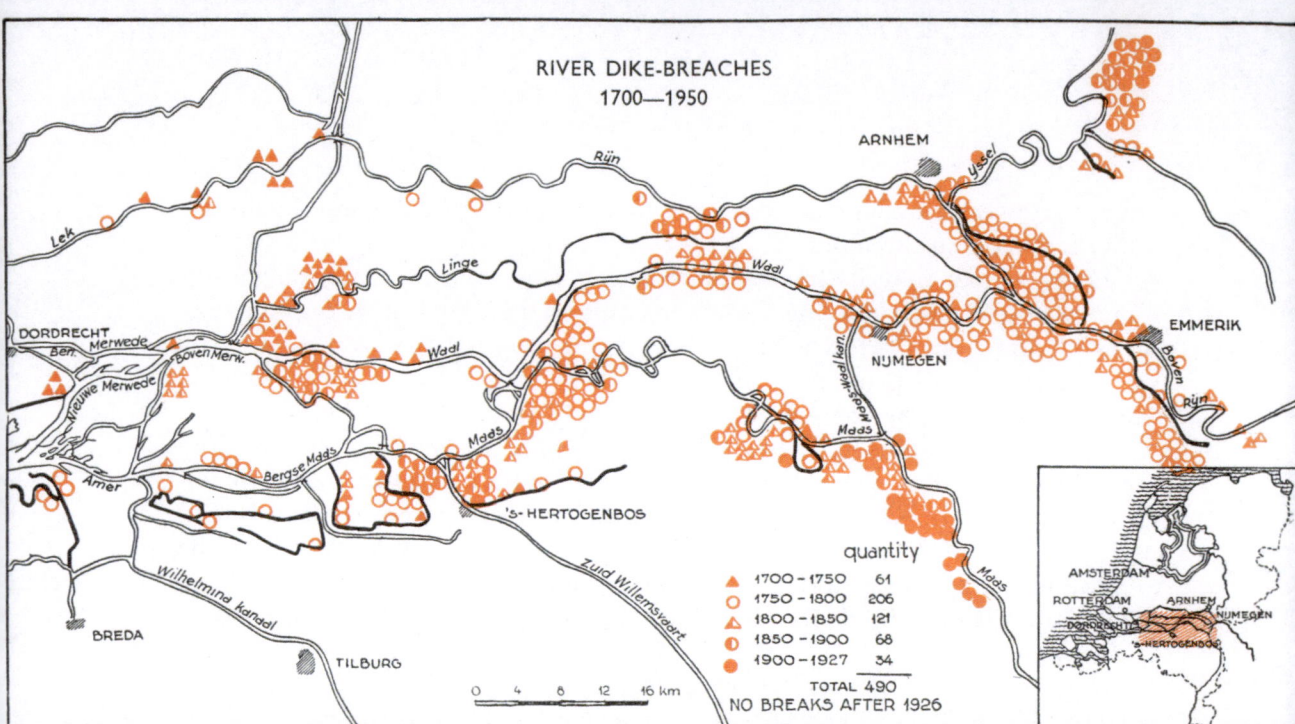

RIVER DIKE-BREACHES
1700—1950

quantity		
▲	1700 – 1750	61
○	1750 – 1800	206
△	1800 – 1850	121
◑	1850 – 1900	68
●	1900 – 1927	34
	TOTAL	490
	NO BREAKS AFTER 1926	

Since 1700 almost 500 breaks have occurred in the main river dikes. The period 1750–1800 was the worst with 4 breaks a year. In February 1953 the breaks in the dikes of the down stream polders were not due to the overflowing rivers, but to the water being driven in from the seaside through the simultaneous action of spring tide and a heavy northwestern storm.

servant, not the master of life. The right to exist comes first. Here is another generally accepted proverb which will endure so long as grass and water endure: *Safety first.* Yes, even before economy. Economists must see that life and its safety are the basic substances of all economy. They should not be called mere imponderabilia.

After the 1953 flood a book has lain constantly on my desk. It contains a compilation, as far as possible, of all known sea floods in Holland. Since 1200 there are records of about 130. Some caused much larger areas to be flooded than did the recent flood; in 1825 the area was two and a half times larger. Moreover there have been hundreds of river floods. *This terror of the centuries must end.* Our modern equipment suffices to make it cease. We are not wealthy enough to suffer such tremendous economic losses over and over again.

Economy, so helpful in many ways, please do not spoil our Ship for a ha'p'orth of tar. The whole of the Delta works now under construction can be had for one year's army budget, a mere trifle in the State-economy of centuries. Four centuries of insufficient upkeep have made the hull rusty. Help save our Ship of State!

27. *Vierlingh's View*

There are three other books about dike building to which I often have to revert: Vierlingh (1575) for the Delta district, Ypey (about 1775) for Friesland,

186

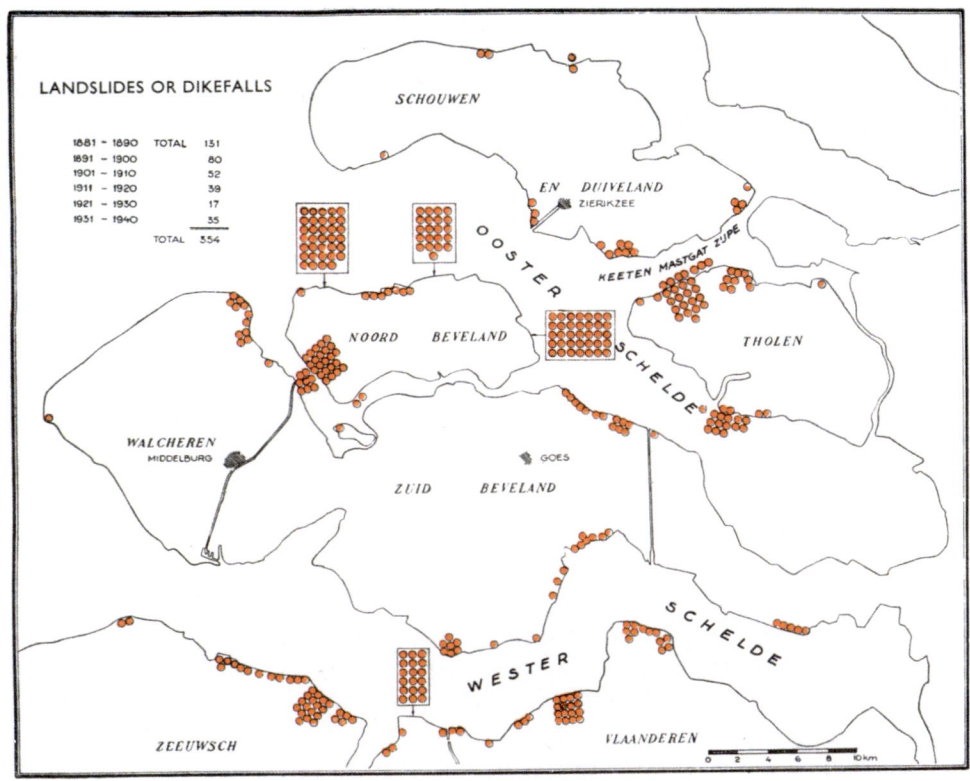

LANDSLIDES OR DIKEFALLS

1881 – 1890	TOTAL	151
1891 – 1900		80
1901 – 1910		52
1911 – 1920		39
1921 – 1930		17
1931 – 1940		35
	TOTAL	354

The above map shows the constant threat of landslides. Zeeland's tidal waters scour its shores in such a way that landslides occur, even though these shores are defended against scour. Viewed over centuries large areas have been lost in Zeeland by landslides, and still new ones are to be feared.

and Abraham Caland (about 1820) for Zeeland. Vierlingh is the most interesting. What does the grand old expert say about floods, about strangling 'roaring' breaches, and about economic values of land?

One of Vierlingh's main fears concerns the island of Schouwen, in 1953 the most afflicted island in the Zeeland archipelago. Almost half of that island had already been lost when he wrote his book in 1575. Landslides due to fierce currents along its shores had already then taken their toll, and great action was necessary in Vierlingh's time to prevent the island from disappearing entirely. Vierlingh estimated that the upkeep of Schouwen would be 'X, yea XX times its worth', and instead of showing pity, he addresses the islanders as follows: 'It is your own fault, you have not withstood the sea in the beginning as the people of the island of Walcheren did. The more you retreat the more the sea has vantage points to throw you out'. – Apparently the Schouwen people clenched their fists after that, for the island still exists. But mind! It was they who had to pay for this upkeep, 10 or 20 times the worth of their land; not the

187

State. The re-claiming of the islands after the flood of 1953 was carried out by the Netherlands State.

Vierlingh never thinks about abandoning inundated lands. He simply says: 'Diking against Neptune and his consorts is a war and we must be warlike'. He dryly says of those communities who have actually lost their land: 'They have not opened their eyes until they had the rod on their backs'. – 'They had forgotten that dikes need a roomy purse.' – 'They forgot that diking is a great hazard and that they never should have started making new land unless they were sure of an annual profit of 10 to 12%.' And he adds, never failing to attack one of the Dutch weaknesses: '*Don't be too stingy and miserly in your dikings*'.

We remember his Master, William the Silent, admonishing the Dutch to take good care of the dikes lest later generations should blame their ancestors. Vierlingh himself hammers: 'Heighten your dikes every seven years, you who yourself have never seen Neptune grim and ugly'. – 'Never sleep, because your foe Oceanus never sleeps, by day nor by night. He comes as a roaring lion to destroy everything; there is nothing on your country that cannot be destroyed.'

Vierlingh counts the heavy land-losses of his days one by one, those lands which could not be re-conquered. He is not impressed by them, he does not believe in breaches that cannot be closed, he says that all the breaches which (in 1575) are really irreparable, could have been closed if people had not quarrelled and if they had been more active in the beginning. The technique of 1575 was advanced enough, in Vierlingh's opinion, only the people were short-sighted. They wanted to learn in the hard way: 'who does not hear must feel'.

Yet, the people of the tremendous floods of 1421, 1508, 1532, 1552 and 1570 had nothing but sailing craft to bring the thousands of acres of mattresses to their places in the strong currents. There are some old maps of the breaches of this time, they show about the same damage as ours; not much was left of those dikes, nor of ours. But we have tugs and the State's purse and power. The people of 1575 were without these. We – and are we weaker creatures? – often wonder, especially when we see that the dikes of 1575 were so much higher and heavier than ours. They seem to have shrunk since then and they have not been heightened 'every seven years' as Vierlingh advised.

28.   The Delta Plan provides for the construction of four main barrier dams, which together will drastically shorten the coastline by about 435 miles. From south to north the following sea arms will be closed: Veerse Gat (1961), Haringvliet (1968), Brouwershavense Gat (1970) and Eastern Scheldt (1978).

This sequence was chosen after due consideration, since the transition from small to large sea arms enables to gain experience to be profitably used in the larger projects. Another reason for this particular sequence is the desire to achieve a higher degree of safety for the largest possible area in the shortest possible

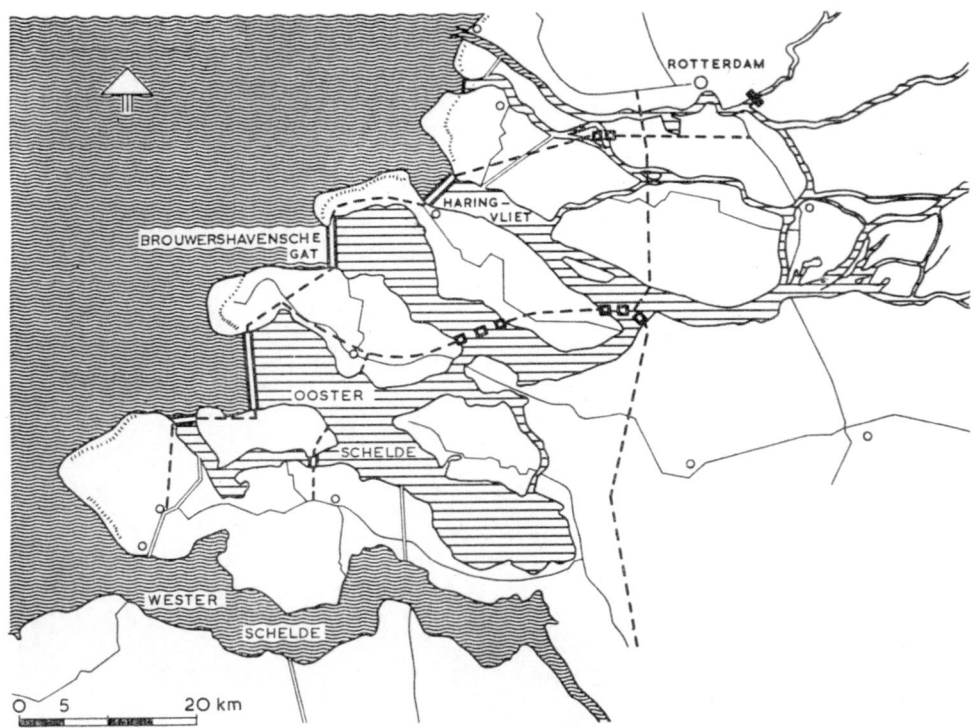

The Delta scheme for closing four southwestern estuaries. A fresh-water lake will be created here. The work will take about 25 years. Total length of dams 25 miles, maximum depth about 100 ft, bottoms consist of loose sand.

time. Although all the works of the Delta Plan must be regarded as parts of a single project, it is true to say that by giving the closing of the Haringvliet the greatest priority, the primary Delta advantages – increased safety and control of fresh water – can be achieved much earlier for the areas between the tidal branches of the rivers Rhine and Maas. Moreover three secondary dams are planned to facilitate the intermediate stages of the execution by preventing currents from gaining inadmissable speeds in the sea arms to be closed off. The first of these dams – in the Zandkreek – was closed already in 1960; the other two – in the Grevelingen and in the Volkerak – are under construction.

In the beginning of January 1954 a start was made with the first Delta work, the enclosure of the river Hollandse IJssel near its mouth, some six miles east of Rotterdam. We have seen before that just here a very dangerous situation during the long lasting stormflood of February 1st 1953 had developed. Especially the dikes along this tidal river have for years been a source of worry to the experts. They are too low to cope with the exceptionally high water levels such as occurred during the flood disaster.

During this emergency it was only a matter of centimetres or the dike

189

The latch on Holland's front door. This movable storm flood defence in the Hollandse IJssel was Delta work number one. In the above picture it is seen from the sea side; below the gate in closed position.

which has to protect the lowest parts of Holland-Proper with almost 4 million inhabitants, would have been breached.

Here the problem could not be solved by simply building a dam in the river, one reason being that the very busy shipping on the Hollandse IJssel could not be interrupted, while furthermore allowance had to be made for the interests of a number of shipyards located further upstream. Hence the enclosure envisaged could not be one of a permanent nature but had to be movable. Next to the storm-flood defence there is an amply dimensioned system of locks, serving a double purpose. In the first place these locks enable shipping to

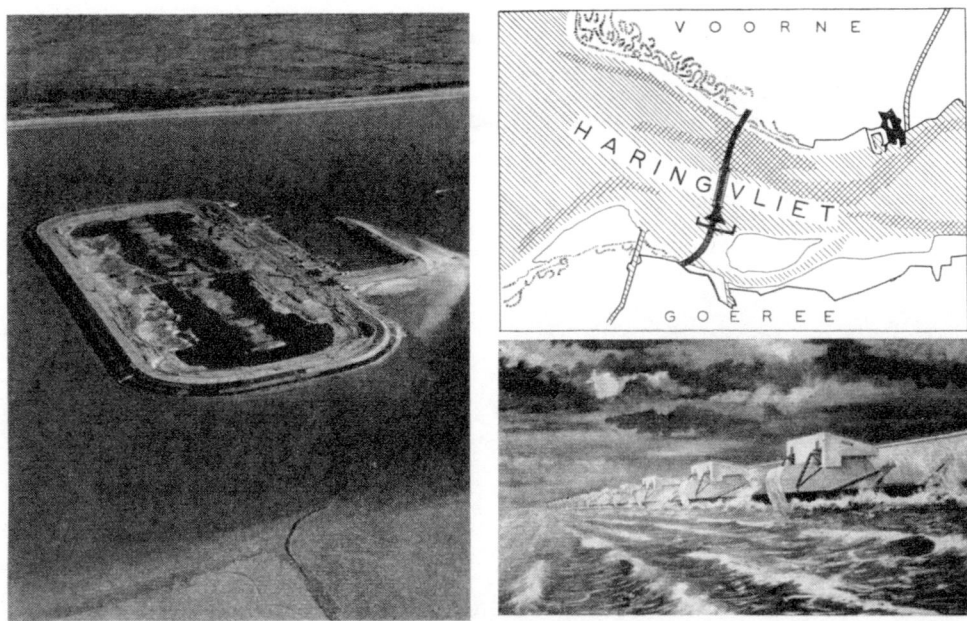

Closing off the Haringvliet estuary. Left the gigantic pit for the seventeen discharging sluices under construction. Right on top the situation, below the planned complex of sluices.

proceed as long as conditions permit when the storm gates have been lowered. In the second place the locks are required to permit the passage of ships with a high superstructure and of floating cranes which would not be able to pass underneath the raised gates.

As a real multi-purpose project the combined stormflood defence and locks have made it possible to create a permanent means of communication with the area called Krimpenerwaard; this polder enclosed by rivers has been an island since ages.

The Haringvlietdam was given, as we have seen already, the greatest priority. It is doubtless the most important estuary in the Delta area. In the present situation more than half of the water furnished by the Upper Rhine flows uselessly to the sea. This situation must be changed, since – as strange as it may seem in a country so rich in water – there is a growing shortage of fresh water every year. One look at the northern part of the Delta area will make it clear that a considerably higher degree of safety and a much more efficient fresh-water economy can be achieved by the barrier dam across the Haringvliet, provided the connection to the south – the Volkerak – is also closed off. When these works have been completed it means that all the outlets of the Rhine and

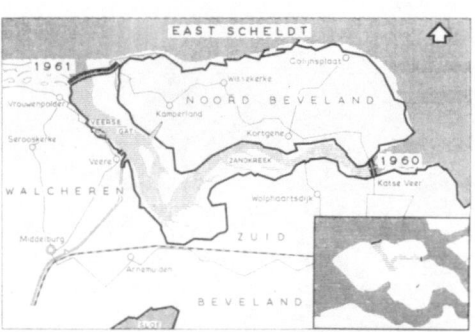

The Three-Island Plan (top right) comprises two closures. The dam in the Zandkreek has been completed. The closure of the primary dam in the Veerse Gat has been carried out by means of so-called water passing caissons (right below).

Maas but one – the New Waterway – will have been provided with movable water defences.

The Haringvlietdam – partly on account of its unusual construction – constitutes a very interesting overture for this unique hydraulic engineering project. Although the problems presented by design and execution should certainly not be underestimated, they are yet of an entirely different nature than those that will be encountered with the dams in the Brouwershavense Gat and in the Eastern Scheldt. The average depth in the mouth of the Haringvliet is less than that of the two other sea arms at the points where the dams will be located. The fact that precisely this dam will be provided with a large complex of sluices with an overall length of more than two thirds of a mile, is a direct result of the circumstance that here it is not only necessary to close off the sea, but that in addition suitable measures must be taken to ensure that in times of very high river discharges a surplus of fresh water, which can no longer be safely stored, can be discharged into the sea.

At the moment two building pits have been erected in the mouth of the Haringvliet; the first one is meant for the construction of the discharging sluices – now in full progress – and the second one is for the construction of a ship lock.

New methods and new materials had to be proved for the execution of the Delta Plan. This floating mechanical way of plunging masses of stones proved to be a success.

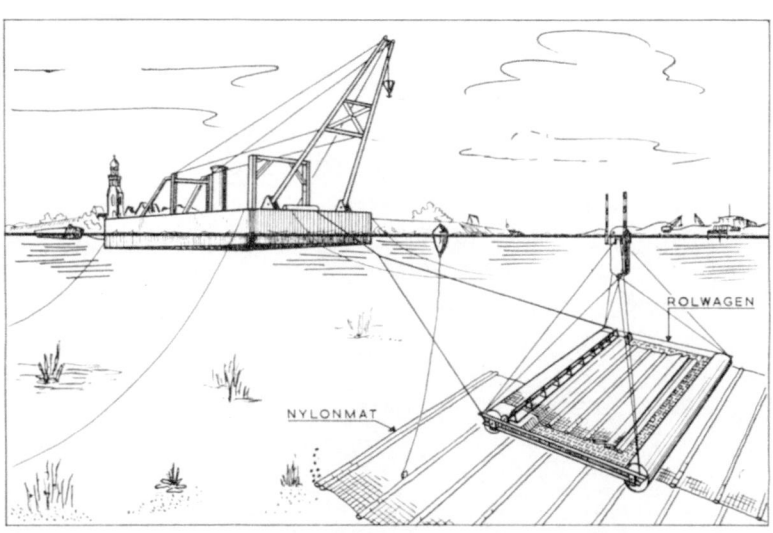

Complete new way of covering a shifty sand bottom with nylon sheets.

Both building pits are temporarily connected by means of Bailey bridges with the coast of Goeree-Overflakkee. According to plan this dam will be closed in 1968.

The object of the so-called Three Island Plan was to join the island of North Beveland by two dams with Walcheren and South Beveland. This project was fully justified in itself, but at the same time it formed a logical part of the much more comprehensive Delta Plan. From the point of view of hydraulic engineering it is particularly fortunate that the shortest of the four main dams was the keystone of the Three Island Plan. It provided the opportunity during the past few years to acquire experience in building a dam in a location directly affected by the open sea. It is, moreover, the first project for which use has been made of the so-called water-passing caissons (spring 1961). Although the closing of the Veerse Gat in itself was not a simple task, it was the considered opinion of the Delta Department, where the plans have been developed, that in the present state of science and engineering the project was a completely safe enterprise. Yet the engineers have had to cope with a tidal volume of more than 90 million cubic yards at an average rise and fall of more than ten feet. What was considered as an operation just within the limit of technical feasibility at the time of the closing of the breaches in the dikes of Walcheren is now – fifteen years later – within the reach of the Netherlands hydraulic engineers. This in itself demonstrates the rapid evolution in this sphere. Moreover, it should be realized once more that the closing of the Veerse Gat is only an initial stage of a project of far greater size.

The crown of this first Delta dam in the frontline against the sea reaches nearly 15 yards above N(ormal) A(msterdam) L(evel), the maximum height of the Enclosing Dam of the Zuiderzee being nearly 9 yards above N.A.L. The dam in the Veerse Gat has been built up mainly with sand. Both on the inside and on the outside the dam has been coated with gravel bitumen aggregate which was covered afterwards with a layer of bitumen concrete.

New methods have succesfully been introduced in hydraulic engineering. It proved to be possible to cover the shifty sand bottom with nylon sheets instead of the classical willow mattresses. An effort to plunge masses of stone in the gullies to be closed was at once successful. Bitumen has been used for the first time below the waterlevel; probably this too is a promise. Only a few examples are given here but it will be sufficient to demonstrate that hydraulic engineering is discovering new ways to fulfil the heavy tasks of to-day and of the near future.

29. The Road Ahead. Allowing for such geological subsidence as has been experienced in the past – the sea level 'vibrates' between 600 feet minus to 300 feet above the present sea level and according to modern investigations based on

194

radio-active peat layers, has risen 55 feet in 7200 years – a short but well-defended coastline could hold the sea at bay for perhaps another thousand years, but the salt, creeping underneath that strong coastal defence, would make the low country infertile. So, after a few centuries, our tools should perhaps be such that the Netherlands could be heightened by pumping sand into them at no unduly great cost. This would be a great task, about 600,000 millions cubic yards for a layer of 30 yards thickness, or about 2000 modern Suez Canals. It would not be a difficult one. There is enough sand and clay in the North Sea and we could start in a slow way, working at an initial rate of say 200 millions cubic yards per year. We could do that now if we wanted to. No wild-horsiness in this, as would say Stevin, nor technical difficulties to speak of.

The most difficult engineering undertaking of the whole Dutch water history, past and future, confronts the engineers of the present day. There cannot be a more difficult problem in engineering construction in all Holland than that which this new challenge presents. It is beset with difficulties because we are concerned with the largest estuaries in Holland. Future engineers will be envious of us in this task. We do not envy the future engineers. Pumping a thick layer of sand and silt into the Netherlands, burying all its history as it were, must be a dull task compared to the present, which is a fighting one against the tides.

Ultimately the Netherlands may not be a pit any longer; it may become an ordinary country, artificial it is true, but above sea level. We cannot wait for the two or three thousand years which are necessary to fill up that pit.

Safety, the first thing, can now be obtained within about 15–20 years; sufficient, preliminary safety. Afterwards the sand pumping might commence. The Dutch interests in the North Sea, the place where the sand and the clay must come from, should be safeguarded, as well as the protection of the Rhine from its progressive salting, which even the German Director of the Rhine describes as 'alarming'. The Rhine should not be a sewer for 50 million people. Without fresh Rhine water the fertility of the Netherlands would be practically nil. Salinity and pollution of the Rhine has already reached such a limit, that crops and recreation are severely endangered. Salmon and other fish have gone in the last few decades.

Will the Dutch nation actually continue its struggle for existence even in a faraway future? It seems largely to depend on the preservation of its fighting spirit, on its continuing to acknowledge that safety comes before economy, and on its willingness to spend part of the annual income for the sake of its progeny; that is to live up to the Zuiderzee motto, slightly altered: 'A nation that lives works for its future generations and their future safety, health and wealth'.

Happy the nation that has challenges and knows how to meet them.

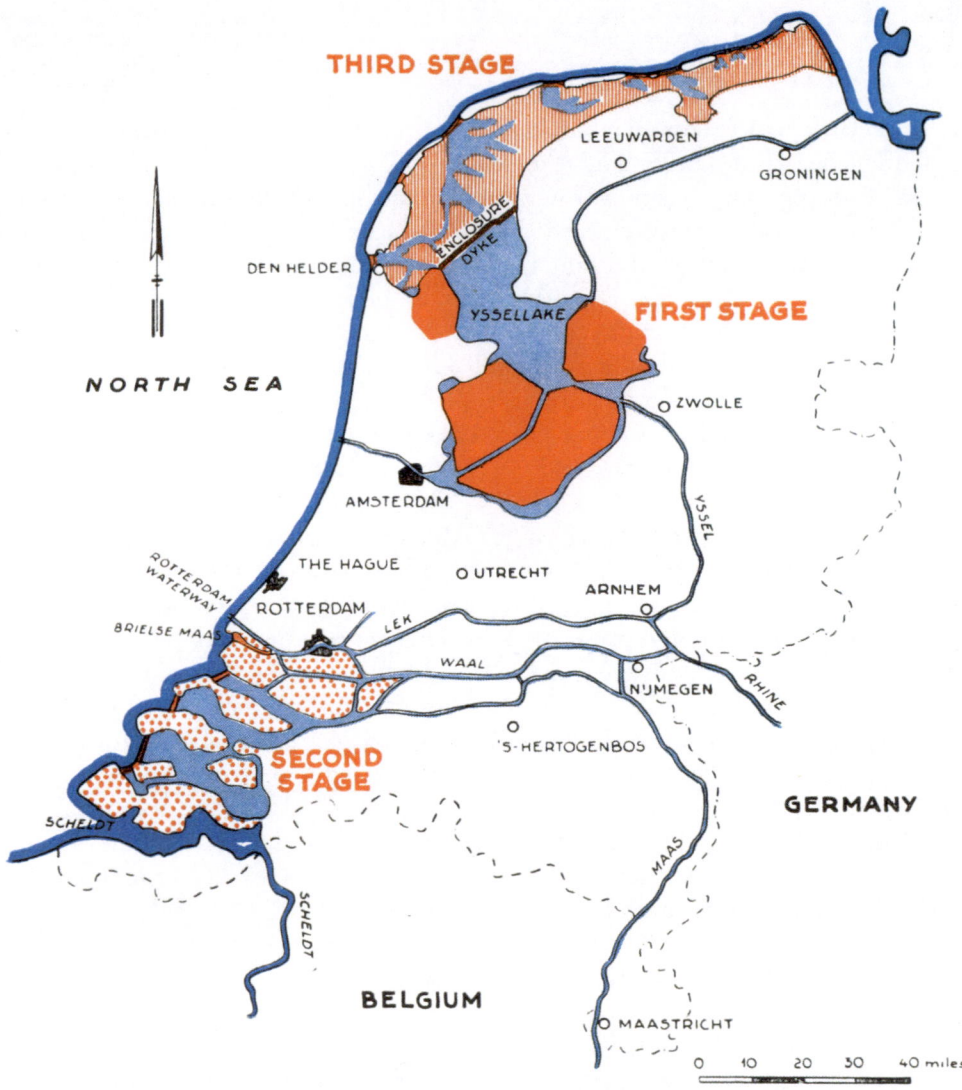

The closing of the Dutch coast will be carried out in three stages mainly. The first work, the Zuiderzee reclamation (1925–1980), the second work or Delta scheme (1954–1978), the third work or Wadden reclamation (not yet dated). The mouths of the Rotterdam and Antwerp Waterways remain open, according to the latest plans.

30. *Luctor et Emergo*. This ancient device of Zeeland spurs us to struggle. We would prefer peace, but the enemy does not want peace, it remains aggressive. And perhaps this is a good thing as it keeps us alert and fit, able to compete.

Catastrophes teach lessons. We have had 140 in the past 850 years, perhaps we shall need more of these lessons, 150 or 170, who shall say? It all depends on our capacity to learn from past mistakes.

We find from our historical studies that some generations have had severe flood-lessons, others not, and that it is difficult to transmit the experience from

196

one generation to the other. There may be such a thing as race-memory, but it is not very active; perhaps it could be quickened by making young engineers study the historical background of the cause of floods: neglect, carelessness, stinginess, and above all lack of organization and co-ordination. – The first thing to do would be *not* to blame the enemy. We should not hear after each flood: 'Our dikes were in a very good shape, yet they broke'. Dikes that break *are* bad, and it is not the hard wind nor the dikes which we can blame. However, it takes some swallowing to say that before the dikes have failed, we ourselves have failed. There is no excuse, now our horsepower capacity has increased a hundredfold since 1575.

Floods can be mastered, we have proved that over and over again, but we now should say: *Floods can be forestalled*. Mastering of floods is technical prowess, forestalling floods is technical plus governmental prowess. It is a kind of cold war, this prevention of floods; may the Dutch be successful by being wise and alert in their Government.

31. Our forgotten six Commandments

In order to be successful in the long term struggle for life, emerging above untoward conditions as the Zeeland device wants us to do, one may not forget the principles of life and government. Let me finish by telling our forgotten settler's saga and the serious warnings *Ald* gave us when we first came here 24 centuries ago. Have we heeded those warnings sufficiently in the past? Shall we heed them enough in the future? We should not have forgotten Ald, our first King, nor his commandments.

When we clear the saga from some of the cobwebs and draperies of so many ages we find the following. The immigrants came by ship, a huge vessel, called 'Mannigfald', which means 'Manifold'. Others call her 'Ship van Ternuten'. The people on board were not at ease of course, not sure of the future now that they had left their home country. They even quarrelled and the strongest party threw some of the weakest party overboard. After these extravagances, during a storm, and with consciences not too quiet, something grey crept on board by way of the bow. It was a Spanking Spirit. It disappeared in the bowels of the great ship Manifold! – Wondering what would happen, the Frisians, after several moments of silence, heard sighings, moanings, groanings, prayers for the future of the race, beseechings, sobs, appealings, clamourings to the people, all very terrible to hear, . . . and after that the words, howled with an enormous Voice through the ship's holds: 'Righteousness, Unity and Hope', repeated a thousand times, until the emigrants could stand it no longer. 'A people cannot exist unless they allow Righteousness, Unity and Hope to live in the midst of them'. Ald (= the Old) taught them this with his booming Voice over and over again. No escape from his terrible but impressive sermon on board the Manifold.

After three days and nights of constant howling and admonishing 'The Old' left as he had come by way of the bow. Nobody had ever seen much of him. Later, in a hidden place at the bow, called Spintje (?), three golden figures were

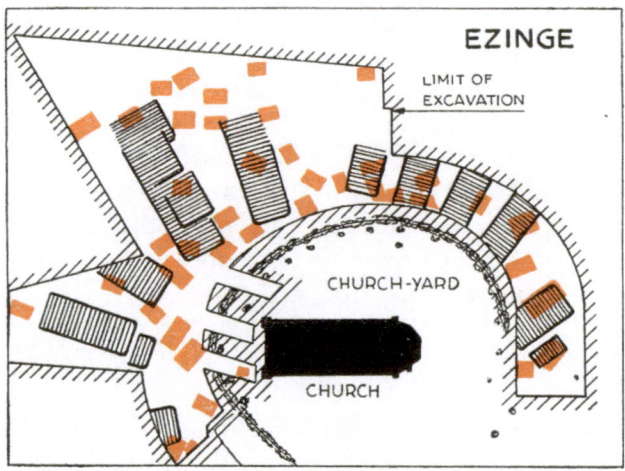

The passing of the Anglo-Saxons about 400 A.D. along the Dutch northern shores according to archaeological excavation of artificial mounds (Prof. van Giffen). The original Frisian farms were standing in a circle on the mound (black hatching). They were burned by the Anglo-Saxon invaders who built their much smaller huts in large numbers without any general planning (red.) After a long time they left or were assimilated, and the Frisian type of farm was built again. The present church (black) is of more recent date, standing about 12 feet higher than the level at the time of the Ango-Saxons.

found, wrapped up in a goatskin: *Righteousness*, a woman with a sword and weighing scales, *Unity*, a mother with three children, and *Hope*, a woman with an anchor and a bird.

For many centuries people carved these symbols on their ships, chests and knife hilts. This custom seems to have ceased, which is a pity. In reality Ald was our first King, the saga says. The common sea-term for captain is still 'the Old'.

We are now on our leaky ship called the Netherlands: a wide and flat-bottomed ship that has hardly any freeboard. Her people are manifold and they still need hope for the future. They have just weathered a storm. What would pre-Christian Ald say, were he to appear again in the hull of our present Ship of State? I suppose he would behave much in the same manner, saying the same words. I almost can near his Voice resounding through the hollow Holland polders. We know that Righteousness, Unity and Hope are necessary and we would say too readily and easily 'yes' or 'amen' to it. But there is more!

Apart from the three commandments mentioned, The Old gave three others.

198

These were written in runic script on the goatskin (parchment). Altogether there were *six commandments*. Mark the word 'choose' in each of the last three ones, it is the key-word of freedom. Did we obey those other commandments?

'When you have left this ship Manifold and when you will have settled into the new land, and if you want to become a 'rechtschapen' (rightshapen = shaped to the perfect example), 'eensgezind (one in aim) and 'lucky' (= happy) people, you shall also have to obey the following commandments:

1. choose your rights (the meaning may be: take good care of the preservation of your rights, find out for what rights you would fight, or: make Right the first principle),

2. in peace-time choose a body of Wisemen who can give you good counsel, choose Righters (those who make crooked things straight), create adequate forms for living together,

3. when trekking or when in danger choose yourself one King, and obey his orders.'

It is evident that we have not obeyed this last commandment. Ald, of course, could not foresee the danger the rising sea level would give when, 'after sailing around all the islands of the world, in quest of good, uninhabited land', he brought us to the Low Lands. We do not blame him that he brought us here, but a very powerful enemy has threatened and killed Ald's race since the beginning; the movements of that enemy are so slow that they cannot be seen in a single generation. We are disobeying Ald because we have not chosen a responsible organization against this danger. Instead the Delta Committee of 1953 stated the situation as follows:

'At present administration and maintenance of the dike system in south west Holland is spread over 3 provinces, 28 communes, 13 'High Dike Reeve Councils', 64 'Dike Reeve Councils' and 125 polders'. (This is only for the south-west!)

No unity in our defence against the sea, no Central Care! What is needed is a responsible Power, stronger than King Canute's, who, as everybody knows, could not stop the rising tide. The time of central vigil in dike affairs seems to have ended in 1581. The Dutch do have a central authority in war affairs against human enemies, but not in the perpetual war against the sea, not even a co-ordinated one. The Government virtually wields no watch and ward in preventing the country from being wiped off the map. 'The Old Man', that forgotten but nevertheless eternal chastising and admonishing character, would certainly howl with indignation for more than three days and three nights.

Of the three golden images, the symbol of the mother protecting her children, should be the symbol of Waterstaat, because this relatively young department was meant to be the central power supervising the ancient lower powers, whose task it is to safeguard the nation against local floods. It should accept its principal function from now on: the co-ordination and supervision of the forces working for the safety against the sea. Shame if it does not, – shame and disaster.

When after 25 years the short, great, new dike on the very shores of the North Sea has been built, let a statue be erected on that dike. To the right: watchful Righteousness with her emblems (the scales weighing the amount of money which should be spent annually on our Safety, and the sword to defend our existence against the North Sea). To the left: staunch Hope with her emblems (the anchor of Security, and the dove of Peace, without which no important building for the safety of the future can be taken in hand); and in the centre the glorious figure of the Mother caring for the safety of the nation in a natural economic plus super-economic way.

Behind this group a towering, massive central figure of *Jan Stavast*[1], our co-ordinated Defence, looking quietly, alertly, intelligently, confidently over the sea into a very far future. Its emblems should be those of resourcefulness, perhaps those of research: a key in one hand and the Box of Unknown Wonders in the other. The device of this immovable and solid figure should be the device of our Queen and her House:

Je Maintiendrai, I shall maintain.

---

[1] ,,Jan Stavast'' may be considered as a Dutch symbol of steadiness and perseverance. *S.*